THE **wildlife TRUSTS**

GARDEN BIRD BEHAVIOUR

*How to recognize and interpret
everyday bird activities*

ROBERT BURTON

NEW HOLLAND

This edition first published in 2007 by New Holland Publishers (UK) Ltd
London • Cape Town • Sydney • Auckland
www.newhollandpublishers.com

2 4 6 8 10 9 7 5 3 1

Garfield House, 86–88 Edgware Road, London W2 2EA, United Kingdom

80 McKenzie Street, Cape Town 8001, South Africa

14 Aquatic Drive, Frenchs Forest, NSW 2086, Australia

218 Lake Road, Northcote, Auckland, New Zealand

ISBN 10: 1 84537 597 1
ISBN 13: 978 1 84537 597 3

Publishing Manager: Jo Hemmings
Project Editor: Charlotte Judet
Copy Editor: Sylvia Sullivan
Design and Cover Design: Alan Marshall
Index: Christine Shuttleworth
Production: Joan Woodroffe

Front cover: Great Spotted Woodpecker
Back cover: Waxwing
Page 1: Goldfinch
Page 3: Great Tits and Blue Tit
Page 4: Song Thrush (t), House Martins (c), House Sparrow (bl), Robin (br)
Page 5: Treecreeper (t), Carrion Crow (c), Magpie (bl), Great Spotted Woodpecker (br)
Page 6: Great Tits

Contents

THE WILDLIFE TRUSTS

The Wildlife Trusts partnership is the UK's leading voluntary organization working, since 1912, in all areas of nature conservation. We are fortunate to have the support of more than 600,000 members, including some famous household names.

The Wildlife Trusts protects wildlife for the future by managing in excess of 2,500 nature reserves, ranging from woodlands and peat bogs, to heathlands, coastal habitats and wildflower meadows. We campaign tirelessly on behalf of wildlife, including of course the multitude of bird species.

We run thousands of events, including dawn chorus walks and birdwatching activities, and projects for adults and children across the UK. Leicestershire and Rutland Wildlife Trust organizes the British Birdwatching Fair at Rutland Water – now also home to Osprey. The Wildlife Trusts work to influence industry and government and also advise landowners.

As numbers of formerly common species plummet we are urging people from all walks of life to take action, whether supporting conservation organizations in their work for birds, or taking a few small steps such as providing food and water for garden birds.

The Wildlife Trusts manage some of the most important sites in the UK for birds. Whether it is Puffins on Skomer Island, Ospreys at Loch of the Lowes in Scotland and Rutland Water, or Bitterns at Far Ings in Lincolnshire, Wildlife Trusts reserves offer fantastic birdwatching opportunities.

For many people birdwatching begins in the local park or garden, and for both beginners and more experienced birdwatchers, *Garden Bird Behaviour* is an excellent guide to the traits, habits and idiosyncrasies of garden birds – what it is that they are doing and why it is that they do it. Robert Burton's animated text describes all facets of garden bird behaviour, from nest building to eluding enemies. Robert explains how to study birds, describes seasonal patterns of bird behaviour and discusses bird intellegence and the function of the dawn chorus, as well as shedding light on more unusual aspects of bird behaviour, such as sparrow parties and fostering. This is a thought-provoking, practical and highly enjoyable book that will help unravel many of the mysteries of garden bird activity.

The Wildlife Trusts is a registered charity (number 207238). For membership, and other details, please phone The Wildlife Trusts on 0870 0367711 or see www.wildlifetrusts.org

PART 1
The Birds in Your Garden

'Feeding the birds' is a pastime that is known to date back at least to Elizabethan times, but we can imagine people from time immemorial throwing out scraps for the pleasure of seeing birds coming to eat them. The poet Geoffrey Chaucer wrote of the 'tame ruddock' (an old name for the Robin), which suggests that it was already garden favourite. Nowadays the natural world has been driven into corners but gardens are becoming tiny nature reserves which, lumped together, make up large areas where many animals live close to natural lives.

RIGHT AND ABOVE: *Birdtables and bird feeders are good places for observing bird behaviour.*

Enjoying your birds

'This is an age of curiosity. There is a desire to know about the private life of people who are much before the public; and birds, as well as men and women, are the subject of more curious and particular observation than they have ever endured before.'

VISCOUNT GREY THE CHARM OF BIRDS

This statement is as true today as it was in 1927 when Viscount Grey was writing. The media today reveals more about the private lives of people, and animals, than would have been imaginable a few years ago. The popular interest in birds, even the common or garden varieties, is shown by the 2003 MORI poll, conducted for the Royal Horticultural Society, which showed that one half of all gardeners put out food for birds and one third watched the

BELOW: Crops of ripe berries lure fruit-eating birds, like this Redwing, close to the house where they can be watched from the windows.

wildlife in their gardens. Gardens give us a window onto nature on a daily basis and provide unrivalled opportunities for observing the private lives of birds that are 'much before the public'.

Watching birds

I spend more hours watching garden birds than most people. I am lucky because I work at home and, as I write about wildlife for a living, I can persuade myself that gazing out of the window is Work.

Even when not peering around the screen of my PC to watch visitors to an array of feeders, I cannot help catching sight of movement in the garden

beyond, and stop typing to watch. My excuse is that much of my writing is about garden wildlife and this apparent idleness is really research. A newspaper editor once phoned to ask for some urgent copy. 'Look out of your window and write about what you see', he said. I did and almost immediately found a subject.

The pleasure I get from the birds in my garden comes particularly from watching what they are doing and trying to explain why they are doing it. The trick is to ask questions, however trivial. Indeed, seemingly simple questions are often the hardest to answer. It does not matter if your question is not new and that the answer is well-known to ornithologists and can be found in books. There is always an excitement in finding out something for yourself.

Keeping records

In *An Englishman's Year* published in 1948, H. J. Massingham described the pleasure to be gained from simply watching birds and the satisfaction of finding out a little about how they spend their lives. He kept watch on a pair of Blue Tits feeding their young while he was confined to a chair in the garden. 'It was beautiful to see them swoop down… Every motion of these jewelled atomies of life was a flash of certainty… What acuity of the senses and brilliance of execution.' But he did more than enjoy watching their comings and goings; he observed carefully what they were doing. They were commuting between apple trees and nestbox with food for their nestlings.

Massingham had the leisure to record that each tit visited the nest 40 times an hour, so the nestlings were receiving 80 meals an hour from both parents. Given 18 hours' daylight, but offsetting time taken for rests and their own meals, this amounted to 1,500 meals a day. Massingham must have derived

ABOVE: *A House Sparrow with unusual plumage can be studied as an individual to see what it does through the day.*

considerable satisfaction from working this out. Even better, he was able to observe that the tits were preying on the grubs of apple blossom weevils and apple sawflies, two serious pests of his orchard.

Local residents and foreign visitors

There is so much going on that I never tire of looking out of the window. I am doubly lucky that I have a large garden in a rural setting so the possibilities of observing bird behaviour are enhanced by the variety of birds coming into the garden. In the five year period between 1998 and 2003 I have recorded 62 species in my garden. The least expected was a Rose-coloured Starling (an infrequent visitor from the steppes of Central Asia) but the Lesser Spotted Woodpecker, Kingfisher and Grey Wagtail were exciting because they are rarely recorded in gardens. (The exotic Rose-ringed Parakeet was interesting but not surprising because it breeds in this county.) Others like Green Woodpecker, Blackcap, Yellowhammer, Kestrel and Little Owl are fairly unusual as garden birds but I see them almost daily at different seasons.

BELOW: Keep watch for unusual birds. Kingfishers sometimes visit garden ponds in winter when they spread out from their usual haunts.

ABOVE: Spotting an exotic visitor, like this Rose-coloured Starling from Asia, adds excitement to garden birdwatching.

It is always nice to see something new and exciting like the Kingfisher diving into my pond. It was perching on the sundial and twice tried to dive in, but was foiled by the netting that was keeping out dead leaves. This was in winter when Kingfishers are dispersing from their normal haunts. In icy conditions they may turn up anywhere and raid goldfish ponds. Amazingly one was seen hovering in front of hanging suet and another, *in extremis*, ate scraps of bread.

But these are rare occurrences and, although I enjoy seeing the more unusual species in my garden, I learn most about bird behaviour from the everyday visitors. These are the common birds that can be seen in almost anyone's garden: Blue, Great and Long-tailed Tits, Dunnock, Blackbird, Robin, Song Thrush, Magpie, Collared Dove, Starling, Chaffinch and Greenfinch and Wren. Many of them come to the feeders and nest either in my garden or my neighbours' so I have the added delight of watching the intimate details of their family lives.

The most frequent visitors

Everyday visitors to the garden are the main players in this book, which examines the daily lives of common birds. Some birds will therefore get an undue share of attention.

The nesting habits of Great Tits and Blue Tits have been studied intensively because they take up residence so freely in nestboxes. Chaffinches have proved good subjects for the study of song, and Robins and Blackbirds for territorial behaviour. The complex mating habits of the humble Dunnock have provided a fertile field of study and the House Sparrow has been used for studying several features of bird behaviour because it is so adaptable and ready to learn.

This book is not a comprehensive encyclopaedia of garden bird behaviour but concentrates on aspects of bird life that I enjoy watching and trying to understand. It is full of my own observations and those of people who have written to me. As a natural history writer, I believe that if I write about what I have seen and find interesting, there is a good chance it will have been seen by other people and aroused their interest. For instance, I have not described how an egg hatches because you cannot see what is happening unless the egg is hatched artificially in an incubator. And I have not devoted a chapter to the important and fascinating subject of migration because, from the point of view of the garden birdwatcher, there is nothing to note except the regular appearance and disappearance of migrant birds.

A bird's eye view

Richard Jefferies, the Victorian essayist, wrote: 'To understand birds you must try and see things as they see them, not as you see them.' It is, of course, rather difficult to stand back and nod understandingly as a crop of soft fruit is ravaged. The exasperated gardener may imagine something deliberately cunning and spiteful about the way a Blackbird finds its way through the protective netting, but the bird sees only the wherewithal to keep body and soul together. 'A whole system of sentiment and conduct', concluded Jefferies, 'has been invented for birds and animals based entirely on the singular method of attributing to them the plans which might occur to a human being.'

In other words, Jefferies is asking us to remove all traces of anthropomorphism. It can be amusing to look at animals as 'little people' but it does not help the true interpretation of their behaviour. By all means have a laugh at them, but do not jump to conclusions about what they are doing based on our own experience. On the other hand, thinking of them as elaborate robots is to miss the complexities of animal behaviour. The moral is that we

BELOW: Familiar birds can be the most rewarding to observe because they interact with each other.

ABOVE: *It is annoying to see a Blackbird stealing soft fruit, but watch how it solves the problem of getting through the net.*

should enjoy watching the animals that share our property and marvel at their ways. They may be more interesting than we think.

Different Studies

Edward Grey, together with W. H. Hudson, T. A. Coward and other naturalists who were writing in the first half of the 20th century, described their own experiences with birds, sometimes adding observations of other naturalists. Their writing was very personal and therefore necessarily limited, although inspiring to read. I often dip into their books, but more for pleasure than enlightenment. I turn to more recent accounts for information on the habits of familiar birds. After World War II, there came David Lack, the Rev E. A. Armstrong, David Snow and others who studied common birds with the critical eye of scientists. Their very readable monographs on species such as the Robin, Swift,

Wren and Blackbird are scientific studies that revealed the richness of the lives of familiar birds in everyday surroundings. Nowadays there is any number of professional ornithologists studying the fine details of the behaviour of common birds, as well as an army of amateur birdwatchers. They have been revealing some fascinating information because the emphasis has changed from simply describing the habits of bird species to finding out how particular aspects of their behaviour are adaptive; that is, how they confer an advantage on an individual bird and aid its struggle for personal survival and success at breeding.

One change that has come about in recent years is the study of garden birds. Previously writers had relied mostly on studies of common birds that had been undertaken in their 'natural' habitats of woodland and fields. It was assumed that their lives would be the same in gardens but it is now known that this is not always the case. The garden habitat may be significantly different from the countryside in important respects and this affects the behaviour of its inhabitants. Robins like scrubby overgrown places for nesting and rear more young than in neat gardens, whereas Blackbirds have been found to breed less successfully in their original woodland habitat. The advantage for urban Blackbirds may be due to fewer predators, at least where the numbers of cats are not excessive.

The problem is that oddities of behaviour are almost impossible to study because they are unpredictable. So interpretation is equally as difficult. On page 31 I describe how a number of birds have been seen apparently grieving over the dead bodies of their mates. This must be an impossibly rare occurrence that is seen only by accident. Even immorally and illegally slaughtering one member of hundreds of pairs of birds is unlikely to result in a repeat of the observation.

You can get far more out of watching birds if you can tell them apart by sex or age. If you see one Robin following another, it could be a male either seeing another male off his territory or hoping to impress a female. With Blackbirds, there is no doubt what is going on because the male, with black plumage, is distinct from the brown-plumaged female.

Obvious: Chaffinch, Greenfinch, Bullfinch, Linnet, Siskin, Blackbird, Great Tit, Great Spotted Woodpecker, Yellowhammer, House Sparrow, Blackcap, Brambling, Reed Bunting, Redstart.

Not obvious: Goldfinch, Swallow, Starling, Pied Wagtail.

During the summer, the garden is invaded by drab-looking birds. These are juveniles that are initially recognizable by their short tails but many also lack their parents' colours until they moult at the end of summer. Some of the juvenile feathers may be retained and experts can distinguish first-year birds from full adults in the winter population.

ABOVE: *It takes an expert birdwatcher to distinguish the tiny differences between male and female Goldfinches.*

Obvious: Starling, Green and Great Spotted Woodpeckers, Greenfinch, Chaffinch, Robin.

Types of study

There are two ways of studying bird behaviour. Both give satisfaction. One is to see some incident you do not understand and try to find out the explanation. You can do this the hard way by further observation or the easy way by looking it up in a book.

For example, the Collared Dove may not be everyone's garden favourite but has some habits that are worth studying. Sometimes a dove perches on a telephone wire in pouring rain, with one wing raised. If you saw this behaviour enough times you might guess that it was using the rain to wash its plumage. But it is quicker to check in a book and find that it is indeed 'rain-bathing'. A number of birds rain-bathe, an uncommon habit that is worth looking out for, but the dove and pigeon family rain-bathe with one wing raised in the same way as they sunbathe.

The other way of studying behaviour is the opposite way round. You read about a particular trait, then go out and look for it. It comes to life before your eyes and you have the pleasure of knowing what is happening. (Even greater pleasure comes from telling other people what is happening!) The Collared Dove gives another good example. Its display flights can be seen at any time of the year. The bird towers up, clapping its wings and rising almost vertically above the trees, then it circles slowly, gliding on outstretched wings and fanned tail. The object is to demonstrate its ownership of a territory to other doves.

The display is a pretty sight and it is interesting to know its function, but my pleasure at watching it

was further enhanced after I had been reading about the technicalities of bird flight. I started to watch the performing doves carefully. When gliding, a dove spreads its flight feathers and fans its tail to create the greatest possible lifting surface so that it glides buoyantly. It is both eye-catching and allows the dove to circle without losing height. I had read that birds spread their wings and tails to increase the duration of glides but the Collared Dove's display gave me an easily observed and practical demonstration of what I had been reading.

ABOVE: *The aerial territorial display of the male Collared Dove is eye-catching and also shows off the bird's flying skills.*

BELOW: *The age and sex of Great Spotted Woodpeckers are defined by the position of the red on head and neck.*

Spotting the difference

Some birds are easier to study because they come in different colours. Look at male Chaffinches in the breeding season. Their plumage is as gaudy as any tropical bird and a small flock mixing with yellow and green male Greenfinches at a bird-table is a riot of colour. The females are dowdy by comparison in both species. The sexes of some other birds around the bird-table have identical plumages: Blue Tit, Robin, Dunnock, Wren and Nuthatch. The reason for the males' brighter colouring is discussed in Chapter 8, but it is something of a puzzle why the sexes look different in some species but the same in others. The advantage to the birdwatcher is that it is so much easier to watch the details of courtship and nesting behaviour if the sexes can be identified. In several species the juveniles can also be identified for a few weeks after they have left the nest by their generally duller plumage. Last summer I saw a male Great Spotted Woodpecker, identified by the red on his nape, introducing a juvenile, with a red crown, to the peanut feeder. There was no sign of the female, with no red on the head, which sometimes abandons the family before they even leave the nest. Alternatively the family may have split and she could have been living with another fledgling a few gardens away.

Observations can be even more significant if birds are identifiable as individuals. Ornithologists use combinations of coloured rings fitted on the legs for the systematic

study of individual birds but it is sometimes possible to pick out birds with unique characteristics; perhaps a damaged leg or bill or unusual markings. It is not unusual for Blackbirds to have some white feathers in their plumage. As I write, my garden is the home of a Blackbird with a broad white collar around its neck, nicknamed 'Fauntleroy' (after the boy in Frances Hodgson Burnett's Victorian romance who sported a large lace collar on his velvet suit). Fauntleroy has been around for a year. I have watched him establish his territory, acquire a mate and he is now singing as she incubates their eggs.

RIGHT: Blackbirds occasionally have patches of white feathers in their plumage so they can be recognized as individuals.

RING FOR INFORMATION

Once in a while I see a bird at the feeders that has a shiny metal ring on one leg. It is frustrating that I cannot read the number engraved on it. This would tell me where and when the bird was ringed. In all likelihood, it would have been ringed not far away and would not be so exciting as a Brambling from the birch woods of northern Norway or a Starling from eastern Europe. Yet its presence in a garden can still be of ornithological interest. Bird ringing has demonstrated that our gardens are used by many more birds than we think. They are on the move through the day, especially in winter when they are not tied to their nests but are free to wander in search of food. Over 1,000 Greenfinches have been caught in a single garden over the course of two months, yet there were never more than a dozen present at one time. Similarly, a single garden may be visited by 100 or so Blue Tits and nearly as many Blackbirds.

If you find a bird with a ring, whether dead, injured or trapped and needing release, read the ring number and address, write them down and double check. Record the following information:
• Your name and postal address (and e-mail address)
• The species (if known)
• The number of the metal ring, stating address if not British Trust for Ornithology or British Museum. If there are coloured plastic rings, give their colour, position on the leg and whether left or right leg.
• Where the bird was found, including the name of the nearest town or village, county and a grid reference if possible.
• The day, month and year.
• Was the bird dead or was only the ring found? If dead, what was the cause of death if known? How long dead? Was it fresh, long dead or not known?
• If alive, how was it caught, e.g. entered house, tangled in garden netting, stunned hitting window - how was it later released, etc.
Send this information to The BTO, The Nunnery, Thetford, Norfolk IP24 2PU. If the bird is dead, send the ring as well. You will be sent details of when and where the bird was ringed.

The garden birds' year

Goodbye, goodbye to summer! For summer's nearly done;
The garden smiling faintly, Cool breezes in the sun;
Our thrushes now are silent, Our swallows flown away –
But Robin's here, in coat of brown, With ruddy breast-knot gay.
Robin, Robin Redbreast, O Robin dear!
Robin singing sweetly
In the falling of the year.

WILLIAM ALLINGHAM ROBIN REDBREAST

I have been keeping records of the birds visiting my garden for about ten years. They show an annual pattern of activity. Visitors to the feeders vary in species and numbers through the year in a fairly predictable way but there is a greater variety of activity around the garden. There is the cycle of breeding, starting with courtship, then nesting and finally the garden is alive with juveniles learning to lead an independent life. At some point in February, herons start to fly over the garden on slow, deliberate wingbeats. I will then see them every day for many months. Their heronry is in the tops of old oaks half a mile away, where they have nested for as long as anyone can remember. There is a local saying that winter ends when the herons return to the heronry and starts again when the last one leaves. I also notice the ebb and flow of visitors into the garden. Whereas most species are resident throughout the year, the winter months bring in obvious immigrants like Fieldfares and Bramblings but also passers-by like Pied Wagtails and Treecreepers, which I never see in summer. Spring and autumn see occasional visits by summer migrants on the move, such as Chiffchaffs and Lesser Whitethroats.

ABOVE: Lesser Whitethroats may pass through gardens in early autumn but they do not stay. They are waiting to set off on the long flight to winter quarters.

Seasonal patterns

The annual pattern of garden bird life is recorded by the British Trust for Ornithology's Garden BirdWatch project, in which thousands of people, not all of them birdwatchers, note what they see in their gardens on a weekly basis.

The accumulating records are revealing interesting patterns of how different species make use of gardens through the year, and across the country. One very clear result is that even birds like Blue Tits and Blackbirds are less numerous in autumn after breeding has finished. They have moved into the countryside and return towards the end of the year. Others, like the Wren, seem to prefer to remain in the countryside for the winter and only come back in spring, providing their numbers have not been pruned by a severe winter. By contrast, Starlings become more numerous in autumn because our native population is augmented by continental visitors in October and November. Another significant observation is that Goldfinches and Greenfinches move from farmland into gardens to visit feeders as winter progresses and natural food supplies dwindle, but Robins may leave gardens in spring and summer because there are not enough resources to support nesting.

To give a date for the start of a season is to impose human rules on imprecise nature. Seasons do not change on a calendar date but merge gradually from one to the next. Spring may come early one year if warm air floods the country or late if winter fails to release its grip under the influence of northerly winds. There are also three or four weeks' difference between the far south-west of England and the northern isles of Scotland, where summer is shorter and winter longer.

RIGHT: A pair of Jackdaws brings sticks to a hollow tree. Carrying nest material is an easy behaviour to record and compare from year to year.

PHENOLOGY

When I was a boy, my diary had a page for noting when I first saw common flowers, butterflies and frogspawn and when I heard the songs of common birds. Recording these 'first observations of the year', such as The First Cuckoo, is called phenology. Long runs of observations allow comparisons between years and show, for instance, the effect of a mild spring.

Phenology is now being used to study climate change. Some common birds are nesting several days earlier than they were 25 or so years ago, as the result of rising spring temperatures. The laying dates of Robins and Chaffinches are two of the many wildlife indicators used by the Government to monitor climate change.

During this season, the winter visitors – Fieldfares, Redwings, Bramblings and Siskins – return to their northern homes. It is not so easy to record a species' disappearance as its arrival. Its numbers dwindle until it dawns on us that we have not seen one of these birds for some time. The first summer visitors are arriving at about the same time but they are noticed because the appearance of a single individual is a novelty and some of them almost immediately add their voices to the chorus of song that has been building up since the start of spring.

Courtship and territorial behaviour start up again in earnest when the weather improves and some species begin nesting early. There is more activity as migrants pass through the garden – watch out especially for warblers but anything could turn up. The starting date and length of the nesting season are generally related to the period in which a species' principal food supply is abundant. The Robin lays early to take advantage of a flush of caterpillars while the Spotted Flycatcher starts late so it can feed its young on the flies that abound at the height of summer.

ABOVE: The garden becomes full of activity in spring. As the days lengthen, birds sing, court and start nesting.

Spring

Spring is the season of transition and beginning; when the world bursts into life after the winter months. Plants bloom, mammals give birth and birds lay their eggs. The common definition is that spring runs from the spring equinox, March 20th, to the summer solstice, the longest day, June 21st. Yet the solstice is known as Midsummer's Day! For the natural world, the alternative season of February or March to April or May makes more sense.

Summer

St George's Day, April 23rd, is the traditional date for the return of Swallows to England, although nowadays the first ones reach the south coast in mid-March and another month elapses before they get to the north of Scotland. Over most of the country, the 'first Swallows' arrive around the traditional date. These Swallows do not, of course, make a summer, and the main influx comes in May.

The other summer icon, the Cuckoo, is usually heard first in mid-April, sometimes earlier, and it also arrives in numbers during May.

When bird books list breeding seasons, they usually give the date at which egg-laying starts. This is a well-defined event compared with the fleeting and easily overlooked bouts of courtship. These preliminaries to breeding may, as we have seen, commence as early as the previous autumn.

The length of the nesting season and number of clutches laid varies according to species. A species' laying date is partly related to the kind of food the nestlings will receive. Most woodland Robins start to lay in the middle of April and their nestlings (and those of Blue and Great Tits) will be fed on the abundance of caterpillars living on the leaves of trees which peaks in the second half of May. The caterpillars, in turn, will have hatched when the leaf buds burst. Spotted Flycatchers, a species unfortunately in decline and now less frequently seen in gardens, feed their young on flies and other airborne insects. These become abundant later in summer and the flycatchers delay laying until late May or early June. Between one year and the next, the timing of egg-laying is refined by the temperature. In most years Great and Blue Tits lay in the first week of May but extreme laying dates range over four weeks from mid-April and are correlated with air temperatures.

Some birds, like the tits, crows, woodpeckers, Nuthatch and Starling rarely, if ever, lay more than a single clutch. Their nesting comes to an end in

ABOVE: A summer scene in which many birds are spending long hours searching the garden for food for their families.

mid-summer and I am always rather shocked and saddened by the realization that this is a sign that the year is now heading towards winter again. Other birds lay two, sometimes three, clutches because their main foods remain abundant over a longer period. Blackbirds feed their young on worms as well as caterpillars and they occasionally lay as many as five clutches in damp summers when the soil stays moist and earthworms are easy to gather.

a significant part of the garden population. Then they wander away and the adults also seem to disappear. This is the start of the 'quiet season' when the bustle of the nesting season has ended and the birds recover from the rigours of family life. They cease to sing and go into retreat to moult. A few weeks later, they will reappear and snatches of song are heard again.

September is a good time of the year to keep a close watch on the birds coming into your garden. There may be strangers amongst them — birds that were not observed during the summer. At this time, it is possible to hear the songs or call notes of Chiffchaff and Willow Warbler and occasionally I see them moving among the leaves in search of insects. The Blackcap is another warbler that puts in an appearance, but it never utters a note. The Spotted Flycatcher is a more noticeable visitor when it makes short, dashing flights to catch insects.

None of these birds stays more than a day or two. They

ABOVE: In early autumn independent juveniles can be seen in the garden. Later on passing migrants put in an appearance.

Autumn

Autumn starts in August, which is the month of summer holidays, but the nights are drawing in. Swifts disappear in this month and other summer visitors are trickling south. Nesting is at an end and the garden is filled with young birds that are often distinguishable by their plumage. They come to the feeders with their parents and for a while they form

are migrants that are beginning to move through the country before the start of their long journey to warmer climes. These birds may not be British natives but Scandinavians that have been blown across the North Sea as they headed south on migration. One intriguing possibility is that some are the neighbourhood's young birds exploring possible breeding places where they will settle to nest next year.

Later, there is often an Indian summer, a period of fine weather which tempts us to think that, maybe, the summer is not quite over. Great Tits,

Robins, Wrens and Song Thrushes sing strongly but Blackbirds and Chaffinches are noticeably absent from the chorus. Coupled with the autumn chorus, rival males are chased, females courted and pairs formed. Great Tits, Starlings and House Sparrows start to investigate nestboxes. Then, as the weather deteriorates, these preliminaries to breeding are abandoned, although some pairs keep together for the winter.

Winter

The current warming of the climate is leading to a change in habits. More birds are attempting to nest early. In the winter of 2002/3, a cold spell in early December put paid to autumn nesting attempts but by Christmas the weather had become unusually mild. Tits, Nuthatches, Blackbirds, Woodpigeons and Tawny Owls took advantage of the warmth to start nesting but another cold snap in January ended most of the activity. These attempts were made in artificial 'warm spots', such as shopping centres, parks and suburban gardens.

ABOVE: In winter birds depend on the dwindling natural food supplies available. Frost and snow bring extra birds into the garden in search of food.

Over the last 40 years the number of Blackcaps overwintering in England has increased. Rather surprisingly they are not the birds that have nested here but immigrants from central Europe. Many become regular visitors to feeders and consume bread, fat, porridge, peanuts and other items that seem unlikely for a member of the insect-eating warbler family.

Winter can be a difficult time for birds, although those that come to feeders fare better than those that are too shy or unable to learn to make use of gardens. Short days give little time to stoke up for long cold nights but, providing food is plentiful and a bird can find a sheltered roost, it will survive. Starvation looms if crops of seeds are poor and food is locked up by ice and snow. The critical time comes at the end of winter when natural supplies run short and there is a period of bad weather.

How clever are your birds?

So, when I see this robin now,
Like a red apple on the bough,
And question why he sings so strong,
For love, or for the love of song;
Ah, now there comes this thought unkind,
Born of the knowledge in my mind:
He sings in triumph that last night
He killed his father in a fight.

W.H. DAVIES THE TRUTH

The study of bird behaviour is fraught with difficulty because we cannot ask birds what is going on in their heads. They cannot fill in questionnaires, or lie on a couch and answer probing questions. We can only infer the reason for their behaviour from careful observation of their actions. During my education as a zoologist, the study of animal behaviour was in its infancy and researchers sensibly concentrated on simpler, stereotyped forms of behaviour which were easier to observe and analyse. The result was that it was easy to get the impression that animals, and particularly birds, which were thought to be less intelligent than mammals, were little more than automatons guided by 'blind instinct'. There were some uncomfortable instances of birds seeming to show signs of real intelligence but, as this behaviour could not be 'dissected' and analysed, they were dismissed as curiosities. This was throwing out the baby with the bath water. As the behaviour of birds has been investigated more closely, it has proved to be much richer than was once thought. Knowing a few 'rules' of animal behaviour helps us to appreciate what they are doing.

Intelligence or instinct?

A couple of examples will show the difficulty of understanding bird behaviour. I have watched Goldfinches behaving 'cleverly', by holding down dandelion stalks with one foot so that they can easily strip the seeds. This is not so much a clever trick that the birds have learned as simply an instinctive pattern of behaviour.

Then there was the hen Blackbird that had three youngsters following her. Two could fly strongly, the third could only flutter a short distance. The hen took off from the lawn and flew over a fence. She was followed immediately by the two stronger juveniles, leaving the backward one behind. It watched their departure, then flew to a bush beside the fence, hopped up branch by branch until it was level with the top of the fence. Then it flew over and rejoined the family. This looks much more like intelligent behaviour because the young Blackbird seems to have recognized its limitations by making no attempt

BELOW: A Goldfinch holding down a dandelion stem while it pecks the seeds looks clever but it is behaving instinctively..

to fly over the fence, and then worked out an alternative strategy before it got left behind. Yet we know nothing about the workings of the young Blackbird's mind. Was it really being intelligent or merely giving the appearance of being intelligent?

Measuring intelligence

It is easy to be misled about what is intelligent behaviour. Building a nest looks very clever but it is an instinct with which every bird is born. The central problem is to decide what is meant by intelligence and how to recognize and measure it. One of the best definitions of intelligence I have come across is: 'The ability to recognize a problem, hit upon a solution and quickly take appropriate action'. Or more simply, it is the ability to cope with a novel situation.

But you have to be careful about what constitutes a novel situation. Take the instance of a Rook that was attracted to the remains of the Sunday roast which had been hung under a bird-table. It landed on the table and, with its bill, pulled up a loop of the string and secured it with one foot. Then it pulled up a second loop and secured this with the other foot. The Rook was now able to reach the bone and peck scraps of meat from it. At the same bird-table, but on another occasion, a gull completely failed to reach a suspended lump of fat. Which was the more intelligent bird? The gull knew that the fat was edible but it lacked the physical ability to get at it. The Rook appeared intelligent but members of the crow and tit families are well-known for pulling strings to get food. Using their feet while feeding is part of their normal behaviour. If the gull, which does not use its feet in feeding, had pulled up the string, that would have been real intelligence!

What is instinctive behaviour?

Much of bird behaviour is instinctive. This needs explanation. We say that the driver of a car instinctively brakes to avoid a collision. What we really mean is that he brakes without thinking. Through training and experience, he has learned to make an emergency stop when danger threatens. If a driver were really able to brake instinctively, he would be able to hit the brake pedal the very first time he took

BELOW: How do birds, such as this Great Spotted Woodpecker, learn about food that is not part of their natural diet?

a car onto the road. Instinctive behaviour is, by definition, inherited from previous generations and is stereotyped so that it varies very little between members of a species. At its simplest, the animal reacts to a stimulus with a set, programmed response.

Instinctive behaviour, with predictable responses to given situations, is adequate for guiding birds through their everyday lives. Its simplicity is an advantage for creatures that have brains largely given over to the complexities of controlling flight. It is only when a bird is faced with a novel situation that its reliance on instinct makes it appear stupid. On page 103, I describe how my father showed me that a nest of Yellowhammers would react to his tap by gaping for food. In this example of instinctive behaviour, a simple stimulus – the vibration of the parent bird landing – results in a simple response – gaping – by the nestlings. The sight of the brightly-coloured mouth in turn stimulates and guides the parent to deliver food. It may look clever but it is automatic – 'blind' instinct. One characteristic of

BELOW: 'Blind instinct' can lead a Robin to mistake a bunch of red feathers for the breast of a rival.

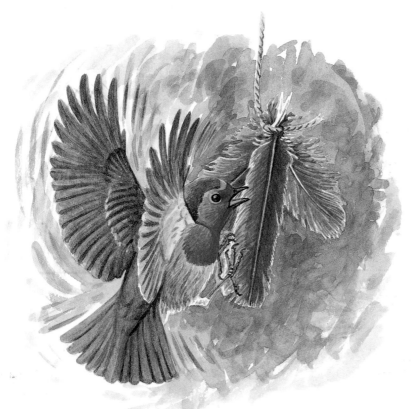

instinctive behaviour is that it can go wrong in unusual circumstances. My father fooled the nestlings into thinking that a parent had arrived with a meal by tapping the nest, but a much more serious mistake was made by two young Blackbirds, newly out of the nest, that gaped trustingly at a cat. They were programmed to react to an approaching object, especially one with a small head on a large body like an adult bird, as if it were bringing food rather than danger.

Many years ago, when open fires were common, I was shown a huge pile of sticks spilling out of a fireplace in an empty house. The pile backed some way up the flue and had a Jackdaw's nest on top. Jackdaws used to nest commonly in chimneys, although their regular sites are hollows in trees and cliffs. The latter places provide a firm foundation but chimneys are bottomless and the Jackdaw's nest-building instinct does not allow for this. In the abnormal circumstances of nesting in a chimney, the birds drop sticks down the flue until one wedges or a pile builds up from the fireplace. Either way it is an inefficient way of building a nest. If the Jackdaws in the old house had been able to think about the problem, they would have carefully wedged a stick or two across the flue to make an initial foundation.

A reliance on instinct does not mean that birds cannot learn and that their instinctive behaviour is completely 'blind' and inflexible. If the bird is to survive in a changing world, it must be able to modify and refine its basic programme of behaviour by learning. Learning is by trial-and-error, or trial-and-success as Konrad Lorenz the pioneer of the study of animal behaviour renamed it. Newly-hatched domestic chicks instinctively peck at anything roughly the size of a seed. At first they are uncoordinated and have difficulty hitting the target. Their aim improves with practice and, as they sample everything they find, they learn by trial-and-error what is edible and what to ignore.

A learning curve

The alternative to learning by practice is learning by observation. I watched a family of Great Tits coming to my peanut feeder. The fledglings settled in the tree where the peanut feeder hangs and the parents pecked out fragments of nuts to take to them. Over the course of two or three weeks the fledgling tits learned to feed themselves. Because peanuts are not one of their natural foods, this seems to be a clear indication that young tits learn where and what to eat by watching their parents. Learning what to eat by observation continues through a tit's life as it watches its fellows to see what they have found.

Birds are making decisions all the time because they are faced with choices: what food to eat; whether to flee or fight; which is the best male to mate with. This does not require intelligence and the bird is using simple 'rules-of-thumb' to reach a quick decision. It may not be perfect but is accurate enough for life to proceed. For instance, when a bird has a choice between two types of food, the rule

LEARNING TO FLY

People talk about birds 'learning to fly' but they do not have that luxury. A bird leaving the nest for the first time must be able to fly in the split second before it hits the ground. It cannot learn to fly gradually, and nestlings reared in confined spaces, like Long-tailed Tits, Wrens and House Martins, hardly get the chance to flap their wings. When they launch themselves into the air as 'learner birds' their rather laboured flight is due to their wing and tail feathers not being fully grown. They instinctively know how to flap their wings to produce lift and have the basics of flight control so they do not crash, but they do need to learn the fine control of take-off and landing through practice.

might be 'choose the biggest items' so it gets the most food for the least effort. Then it might be modified by the corollary 'avoid anything with black-and-yellow stripes', so it avoids being stung by wasps.

BELOW: House Sparrows are very versatile at finding new sources of food. Here they are picking dead insects from a car.

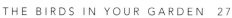

New habits

The standard story of the spread of a new habit is how Blue and Great Tits learned to open milk bottles. This was first recorded near Southampton in 1921 and the habit quickly spread around the country.

It has since been recorded in other European countries and 14 species (including the Great Spotted Woodpecker) are now known to have acquired the trick. It has even been found that, given the choice, tits prefer creamy milk (with plain silver tops) to skimmed milk (blue and silver tops). The explanation of how tits took to opening milk bottles is that it is their natural habit to rip bark to expose insects underneath and there is evidence from experiments that tits with no previous experience will tear the foil on milk bottles. How these new habits are passed from bird to bird through the population is not properly understood. Tits have the capacity to learn by watching other tits, so they have most likely learned from birds that were already stealing milk. Alternatively, a tit could learn to peck bottle tops after it has enjoyed milk from bottles that have already been opened by other tits.

The brainiest birds

More puzzling is how a bird acquires a brand-new habit that is completely out of character. There was once a woman who used to visit Kew Gardens in winter with peanuts for the birds. Robins and tits were the usual beneficiaries but one day a Nuthatch showed interest as it clung to the side of a tree-trunk. It would not come for a proffered peanut so she threw one at it. The Nuthatch flew out and caught it. The trick was repeated on numerous occasions, even when the peanuts were thrown quite hard. Nuthatches usually pick their food from trunks and branches but they have been recorded, on rare

BELOW AND BELOW LEFT: Garden tits are quick to learn new sources of food. New habits spread by birds learning from each other.

occasions, as catching insects in the air. Yet it requires a leap of imagination to intercept flying peanuts, which the bird was unlikely to have encountered even in the stationary state.

Sparrows are particularly good at finding novel ways to get a meal. They have been seen plucking insects from spiders' webs and dropping seeds on to flagstones to split the husks. Some cheeky individuals have entered nests of other sparrows and even a House Martin to take food intended for the nestlings. Another stole from a Blackbird while she was feeding her young on a lawn.

So, although it is the Blue Tit that has earned the title of 'Bird Brain of Britain' for its ability to solve problems, the House Sparrow is a strong contender and there are many stories of sparrows' speed of learning. An American psychologist compared monkeys, white rats and House Sparrows and found that the birds, normally regarded as inferior in brain power, performed as well as the mammals in learning to negotiate a maze to reach food. The learning power of birds is easiest to see in gardens when sparrows and other species learn to deal with bird-feeders by emulating the naturally acrobatic tits. It is very likely that the novices learn by observation and imitation, coupled with a natural curiosity.

Reasoning

Many instances of apparent reasoning by animals have been shown to be due to trial-and-error but there are cases of birds learning something new through insight – the appreciation of the relationships between things, or 'putting two and two together and making four'.

There was a Magpie that pecked at a crust and, finding it too hard, walked to a shallow birdbath. This time it was dry. The bird stood for a few seconds, crust in beak, apparently examining the dish. It then seemed to have a brainwave, flew to a pond at the end of the garden and dipped the bread into the water several times until it was softened. There seems to have been a modicum of thought in the behaviour of this Magpie, but I have been careful in my reporting to employ words such as 'apparently' and 'seem'. I have no idea what was going on in its brain. Even blind instinct sometimes looks remarkably intelligent.

For a bird, a deformed bill can be a matter of life and death. It is not unusual to see Feral Pigeons in city squares with misshapen bills in which the upper half is long and curved. They are unable to cope with seeds and crumbs because they aim their pecks as if the bill were normal. If the pigeon is caught and its bill trimmed, it can pick up food without any difficulty. This would suggest an inability to adapt and a limited intelligence but many of these pigeons survive because they learn to take food from an outstretched hand. Other species have learned to cope with similar deformities, like the Blue Tit with a bill twice the normal length that scooped porridge oats with a sideways motion. These stories are capped by a report from 50 years ago of a Robin that had lost the entire bill. It fed by tilting its head and shovelling food with its tongue.

ABOVE: *Even Pheasants, who are not known to be the most intelligent of birds, can learn how to get peanuts from a birdfeeder.*

RIGHT: *Despite a deformed bill, this Redwing has learned to modify its feeding behaviour and survive.*

In 1941, Maurice Brooks-King decided to test the problem-solving powers of tits by offering them peanuts in 'puzzle boxes' made from matchboxes and mounted on a window frame. The simplest puzzle required a tit to push down the inner part of matchbox to open it and release the nut. In more complex versions it had to pull out matchsticks before it could retrieve the peanut. The response was very variable. Some birds 'got the idea' quickly; others were interested in the nuts but never worked out how to get them.

This puzzle is easy to make and gives hours of amusement to the onlooker and, sometimes, a reward to the tits.

RIGHT: *The matchbox test. The peanut is hidden inside the box and the tit has to push down the inner part of the box to release it.*

British Birds, vol. LI. 1958.

Emotions

There are times when a bird does something that is very unusual and for reasons that are inexplicable.

On at least two occasions, gangs of House Sparrows have been seen slithering down snow-covered roofs on their feet and tails, then flying to the top and repeating the slide again and again. A Blackbird has been seen doing the same thing, except without the snow. Thousands of skiers will testify to the pleasure of sliding in the snow, but can we really transfer our emotions to animals and say that these birds were simply enjoying themselves?

Playing, pleasure and pain

It sounds as if these birds were playing. It is well-known that many animals play. Young mammals, especially, indulge in mock-fighting with their peers or pretending to hunt, sometimes using inanimate objects as 'prey'. This is believed to be part of the learning process and, although it is amusing for us to watch, it is probably serious training for the animal. When it is impossible to think of a function for play, as with the birds sliding down roofs, it is impossible to suggest any other reason than that they are doing it for fun.

The question of whether birds feel pleasure or any other emotion intrigued Charles Darwin. He wrote: 'The lower animals, like man, manifestly feel pleasure and pain, happiness and misery'. For people who live with pets, this is not such a surprising statement, but the idea of animals having emotions has long been shunned by the scientific community as being too anthropomorphic. It was impossible to know what is going on in an animal's mind and, anyway, it was believed that animal brains were probably not sufficiently well-developed for such high mental activity.

Perhaps the most difficult emotion to accept among animals is that of grief – the feeling that someone or something importantly familiar has gone beyond recall. Is it possible for an animal to realize that it has lost for evermore the company of its mate, offspring or close friend? There are numerous instances of birds showing signs of grief when a mate has been killed. They include Swallows, Greenfinches, Rooks, geese and pigeons. Typically, the grieving bird remains near the body of its dead mate and shows signs of what can only be called dejection. My family once kept a male Rook and a female Crow in an aviary. They courted and the Crow laid two eggs, but she died almost immediately afterwards. For two days, the Rook perched in a corner of the aviary, hunched up, not eating. By the third day he had recovered his spirits. We cannot know what went on in his mind, but recent studies indicate that the

ABOVE: *Pleasure seems to be the only possible explanation for House Sparrows repeatedly sliding down a snow-covered roof.*

BELOW: *Swallows are one of several species of birds that have been recorded as showing signs of apparent grief when their mates die.*

parts of the human brain that govern the emotions have parallels in animal brains. They may not experience grief exactly as we know it but they may possess the basics of this emotion.

In the air & on the ground

'Only a small part of their story is revealed by study from the ground, for it is the power of flight which dominates them, and a freedom is theirs which knows no barrier but the wearying of their wings.'

HARALD PENROSE | FLEW WITH THE BIRDS

From earliest times, people have watched birds flying overhead and envied their freedom. Very few have had the pilot Harald Penrose's opportunity to take to the air in a glider or light aircraft and observe birds by flying among them. The power of flight is what makes a bird, yet I am often surprised how books on birds devote so little space to flight. Flying is the key to understanding the lives of birds. Their anatomy and physiology have been radically redesigned to make a strong but light body that produces a prodigious output of power. Almost every aspect of nearly every bird species'

behaviour, from breeding to feeding and including long-distance migration, is based on being able to fly. Flight confers freedom to travel and exploit new resources but it also imposes constraints on birds' activities because it is very costly in terms of energy. One consequence is that birds have been robbed of hands and teeth, which have been replaced by the versatile bill as a tool for handling food.

BELOW: *The power of flight allows Fieldfares to escape cold northern winters and move around the country in search of food.*

The mystery of flight

I have found that trying to interpret bird behaviour in terms of the freedom of flight and the constraint of power economy has added an interesting extra dimension to birdwatching.

There is no need to understand the mechanics of bird flight to appreciate the way birds exploit the airspace. In fact, no one really understands how birds fly because the physics of flapping wings is so much more complex than that of the fixed wings of aircraft. But it is still fascinating to see how birds use their flight skills to exploit the airspace.

One small point has often struck me when birds have flown past close enough for me to hear the noise of their wingbeats. Flying must be like riding a motorbike in which the engine noise drowns out all other sounds. The traffic noise of major roads has been shown to affect birds' ability to hear their neighbours' songs, so it seems possible that their own wingbeats could mask the voices of other birds or the approach of predators. On the other hand, perhaps the rush of wings alerts small birds to a Sparrowhawk coming round the corner.

Air time

There is a big difference in the way that birds use their wings. Pheasants and Moorhens prefer to walk, and fly mainly as a means of escape but, for most birds, flight gives them the ability to search wide areas for food. A Robin feeds within its territory and spends no more than about half an hour per day in flight. Most of this time consists of flying down from a perch to seize a morsel and flying back. The majority of garden birds, such as flocks of tits, finches or Starlings, spend more time in the air and fly around a circuit of feeding sites. This can be seen in the garden when a flock descends on a feeder and then moves on. It is most obvious in the way that parties of Long-tailed Tits stream through the garden, continually on the move. The great advantage is that if one supply fails, perhaps because someone has gone away and the feeders have not been replenished, the birds simply move on to the next site. Conversely, their travels also lead them to discover new feeding places.

ABOVE: *When a bird settles, its legs bend and tendons pull the toes so they clamp to the perch. The grip is so tight, the bird can perch on one leg while retracting the other into the plumage to keep it warm.* BELOW: *A Great Tit approaches a hanging feeder. Understanding the power of flight is the key to understanding bird behaviour.*

ABOVE: Swallows spread their long, forked tails to give extra manoeuvrability when flying at low speed in search of flying insects.

Life in the sky

Swifts, Swallows and martins are the most aerial of birds and they spend most of the day on the wing (perhaps all day for Swifts when not breeding) as they search for airborne insects. Economy is therefore all-important and their long, narrow wings are designed for slow but efficient flight.

Speed is not important for catching flies and Swifts are badly named. They may be swift when they chase each other around the houses on a summer evening but they move slowly when they are feeding. I have watched Swifts feeding at low level and compared them with other birds flying past: there was no noticeable difference in speed. The Swift's real skill lies in its slow sweeping glides in which it efficiently picks up scattered small insects. By contrast, the long, forked tail of a Swallow makes it extremely manoeuvrable and it can 'turn on a sixpence' to seize a passing insect.

BELOW: The Swift spends most of its life in the air except when nesting. It uses very little energy when airborne.

TOO HEAVY TO FLY?

Birds have a weight problem. We think of them gathering on the bird-table on cold winter days to put on a thick layer of fat to keep warm and guard against starvation. However, a plump body is a grave disadvantage because carrying extra weight makes flying harder. The best policy is to keep the weight down during the day and have a good feed in the late afternoon before retiring to roost.

Another drawback is that heavy birds are slower to take-off and are less manoeuvrable. Consequently they are at greater risk from surprise attacks by Sparrowhawks. So putting on weight in the winter is not always the benefit it might seem. Luckily, birds respond to changing conditions very quickly. While the weather remains mild, they keep their weight down, but when the weather is very cold and the nights long, they very quickly put on weight. When the temperature rises, the birds slim down again.

Energetics

A bird in flight uses relatively more energy than a man sprinting. So it is not surprising that birds take steps to fly economically.

The simplest way to save energy when airborne is to glide. When they are not in a hurry, crows, pigeons, gulls, herons and Starlings 'cut their engines' and glide. This is called 'undulating flight' because the bird gains height as it flaps and sinks as it glides. If there is a good wind to carry it, the bird can glide more than it flaps and may make energy savings of up to 80 per cent.

Small birds are poor gliders and their solution to energy-saving is to alternate bursts of flapping and hurtling through the air with wings closed. This is called 'bounding flight' because the bird describes a roller-coaster path, flying upwards, folding its wings and dipping, then flapping and rising again. Bounding flight is used by tits, finches, sparrows and other birds up to the size of woodpeckers and Little Owl. It is very exaggerated in wagtails but watch a flock of finches on a windy day bouncing into the breeze, as if on elastic, and hardly making headway.

Thermal energy

Some birds such as Rooks and Buzzards, use thermals of warm rising air to buoy them aloft. With Rooks the whole flock spirals up, with pairs keeping together, and then descend in wild swoops known as 'shooting'. It is not clear why they do this. It may be a part of courtship, a form of play, or both. Swallows, martins and Swifts employ the upward flow of air along rows of trees, embankments and even sunken lanes, where the turbulence also traps insects for them to catch. I often have a flock of House Martins feeding over my garden in late summer. The flock appears when the wind sweeps across the open fields and swirls over a row of trees.

TOP LEFT: A Green Woodpecker saves energy by 'bounding', alternately flapping and closing its wings.

ABOVE: Flocks of Rooks use the rising air in thermals to climb effortlessly to great heights, then come swooping down.

ABOVE: Hovering is very strenuous but it allows the Kestrel to hunt over open ground where there are no perches.

The martins face into the wind, almost hovering, which keeps them airborne with the minimum of exertion and positions them to catch insects caught in the eddies. Feasting on a static swarm of insects saves any amount of patrolling around open country, especially when the weather turns autumnal and insects are harder to find.

Flight economics

At times, birds must forego economy. They obviously fly as fast as possible when being chased by a hawk and when they are racing to reach food before their competitors.

Sometimes birds hover to get food. Hovering is the most strenuous kind of locomotion in the whole of the Animal Kingdom; most birds can hover for no more than a few seconds. Blackbirds and thrushes hover to pluck berries and a variety of species hover momentarily to pick insects from places where they cannot land. I once kept a record of the birds that hovered at a tit-bell. While the tits nimbly turned upside down and grasped the rim of the bell, less agile birds – House Sparrow, Robin, Chaffinch, Starling, Blackbird, Mistle Thrush and Nuthatch – had to peck the fat while hovering under the bell.

Highway hovering

Kestrels are famous for hovering, often viewed hanging over a roadside verge, waiting to drop on unsuspecting prey. A good breeze, especially one flowing up an embankment or over a line of trees, will be enough for a Kestrel to hang on its outstretched wings. As the wind drops, it has to spend more time beating its wings.

The advantage of hovering is that it enables a Kestrel to extend its hunting range over open ground. But, unless there is a stiff wind to provide lift, hovering is very hard work. It is most worthwhile in summer because prey is sufficiently abundant to give good returns for the effort and the need to feed the nestlings overrides any requirement for economy.

Hovering is so characteristic of Kestrels that it is a sure means of identification but it is not the Kestrel's only technique for hunting. I have been watching one of our local birds searching a paddock where the close-cropped turf must make it easy to spot small prey. Instead of hovering, my Kestrel sat on fence posts or overhanging branches.

Aerial acrobatics

The need to win a mate is another reason for abandoning economy. A male bird spares no effort to achieve fatherhood and one way to demonstrate ownership of a territory and need for a mate is to show off with an aerial display. The best-known is the song-flight of the Skylark. Rather less spectacular are pipits and Whitethroat, although these are unlikely to perform over gardens. The song-flight most commonly seen over gardens is that of the Greenfinch, but look out for the less common but similar Goldfinch song-flight. Greenfinches often sing from the top of a tree, then sometimes take-off and circle the nesting area with a distinctive flight of slow, deep wingbeats, singing as they go.

Take-off and landing

These are the easiest aspects of flight to study. I find the comings and goings of birds endlessly fascinating. Small birds land too quickly to enable one to see precisely what they are doing, but the manoeuvres of crows and pigeons are slow enough to follow.

An aircraft takes off by rolling down the runway, gathering speed until it has enough lift to get airborne. Most birds take off from a standing start. They crouch then spring into the air for a vertical take-off. (Common exceptions are Moorhens and Coots which patter across the water.) Taking off is strenuous, especially for larger birds. I have two feeders on my back lawn set 20 metres apart and small birds feeding under them regularly fly between the two. Heavier Woodpigeons, Stock Doves and crows walk, while Blackbirds are in two minds: they often hop and run for part of the way.

Slow and steady

A Woodpigeon cannot take off from the ground more than twice in the space of two minutes because the effort is so great. It rises with a loud clatter of wings which was once thought to be caused by the wings clapping as they smacked together at the top of the upstroke, but the action is more complex. The wings flap in deep, exaggerated beats to lift the pigeon vertically clear of the ground before speeding away. For the first few strokes, the body is almost upright so the wings are swept forward and back almost horizontally, while the bird has hardly any forward speed. It is rather like the vertical take-off of a helicopter. Extra lift is created by pressing the wings together over the back and whipping them apart, with a loud crack, to suck air over the wings.

Accuracy and control

Small garden birds land more like a helicopter than a fixed-wing aircraft. They slow

down until they have zero forward speed, while flapping hard to maintain as much lift and control as possible. Landing on a perch requires considerable accuracy. Small birds clearly have no problem as they simply pitch in, and acrobatic tits and warblers regularly land upside down. Once again, it is the large birds that are interesting to watch. When a Woodpigeon lands, it alternately furls its wings to drop rapidly, and spreads them to momentarily check its descent. Perhaps this is a way of reducing the time when it is particularly vulnerable to predators while still maintaining control.

ABOVE: A Blackbird crouches with wings raised in preparation for launching itself into the air. BELOW: A Woodpigeon takes off with exaggerated wingbeats that maximise lift.

Birds on the ground

Birds have two gaits on the ground: hopping and walking. Most birds either walk or hop but a few do both.

I have kept watch to see how different species perform on the ground. The Mistle Thrush that visits my garden hops across the lawn, but when it spots an unwary worm or grub, it hurries forward with a rapid-paced walk. Blackbirds, Song Thrushes and crows are runners but they hop when in a real hurry, as when a male Blackbird chases a rival or when crows race to be the first to reach some food.

FORMATION FLYING

One of the greatest spectacles of everyday wildlife, in town or country, is the evening gathering of flocks of Starlings. It is best seen on a clear evening when the flock is flying against a sky glowing orange or pink. Starlings converge from miles around to spend the night in huge communal roosts. For several minutes before they enter the roost, the Starlings fly to and fro in swirling clouds of packed birds. The flock stretches out into a wisp, gathers again into a compact bunch, drops earthward and rears back up, roller-coaster fashion.

That collisions are rare is a testimony to the Starlings' skill at formation flying. It looks as if they are changing direction at a command but each bird is responsible for its own movements. If it reacted as the bird in front of it altered course, its reaction would come too late. To obtain precision aerial evolutions, each bird watches birds several places farther ahead and takes its cue from the first signs of their changing course. In this way, the entire flock wheels smoothly and almost instantaneously. There are parallels in driving on a busy road. A good driver watches the brake lights of cars farther along the queue so there is time to react smoothly.

ABOVE: *Starlings in a flock watch each other carefully to prevent collisions.*

BELOW: *V-formation helps gulls because they fly in each other's slipstreams.*

To hop or not to hop

The explanation for hopping is that it is the best way of moving from twig to twig in a tree and that birds which are more at home in the trees hop when they come to the ground. Birds that habitually feed on the ground, such as wagtails, crows and Skylarks, walk. Greenfinches and Great Tits hop although they often feed on the ground, but Chaffinches, Wrens and Dunnocks, which are even more ground-loving in their habits, hop and walk. There does not seem to be any inherent advantage in either gait for moving on the ground. But it is noticeable that Blackbirds and thrushes end a series of fast hops with a short run, as if they would over-balance and fall flat on their beaks if they simply stopped hopping.

Walking and wagging

Walking birds, from Pheasants to Chaffinches, nod their heads at each step. In effect, they are keeping the head fixed relative to their surroundings while the body keeps moving. This makes it easier to spot small morsels of food or approaching danger. Hopping birds have time to fix their gaze and

scan their surroundings between each hop.

As well as nodding its head, a wagtail wags its tail. Several theories have been put forward by ornithologists to explain this but none seems to fit the bill. I wonder whether there is a parallel with the behaviour of the Australian Willie Wagtail, which looks like, but is not related to, our wagtails. When feeding on insects, the Willie Wagtail wags its tail and also flashes its wings. This is believed to disturb flies, so they fly up and become easier to spot.

ABOVE AND BELOW: Watch birds moving on the ground. Some, like the Starling and Pied Wagtail, walk and nod their heads. Others, like the Robin, hop. A few use both gaits.

PART 2
Their Daily Lives

Feeding is the key to all birds' survival. They must find enough food every day to fuel their activites. Feeding is the behaviour that can be watched most easily in gardens, especially if plenty of food is provided on birdtables and feeders. As well as the pleasure to be gained from watching the colour and movement of visitors to the garden, there is insight into the feeding habits of different species. The Exchanges between the birds also shed light on the secrets of their social lives.

ABOVE: Goldfinches are great favourites when small flocks visit teasel heads. *RIGHT:* The Waxwing is an uncommon but striking winter visitor that specializes in eating fruit.

Conspicuous consumption

'The blue titmouse, or nun, is a great frequenter of houses, and a general devourer. Besides insects, it is very fond of flesh; for it frequently picks bones on dunghills: it is a vast admirer of suet, and haunts butchers' shops… It will also pick holes in apples left on the ground, and be well entertained with the seeds on the head of a sunflower.'

GILBERT WHITE THE NATURAL HISTORY OF SELBORNE

Birds that come into the garden to feed face an artificial situation. Gardens do not bear much resemblance to the woods, hedgerows and fields that have been the homes of our common birds for thousands of years. One average-sized garden is too small a plot of land to contain everything a bird needs so it has to forage around the neighbours' gardens. Furthermore, a garden is a simple, tamed habitat modified by planting, pruning, and removing weeds and insect pests. Consequently there is

BELOW: Birds' bills come in various shapes and sizes. They give an indication of what food is likely to be eaten.

Goldcrest

Song Thrush

Greenfinch

Great Spotted
Woodpecker

less natural food for birds. And we attract them with unnatural food such as peanuts and scraps. However, this does not stop us from watching some of our common birds' natural feeding habits.

Bills and diets

I had a good opportunity to make some observations on feeding behaviour when my cherry tree bore a heavy crop. It had become a major attraction for Blackbirds and Starlings, so much so that the birds were stripping the cherries before they had ripened.

I was interested to see the striking difference in the way that the two species dealt with the cherries. The Blackbirds plucked and swallowed them whole but the Starlings pecked off the flesh and left the bare stones hanging on their stalks.

The difference in behaviour lay in the shape of the birds' bills. The Starling has a narrow bill designed for probing into the ground for insects, while the Blackbird has a proper fruit-eater's bill with a slightly hooked tip for plucking fruits and a broad base to give a wide 'gape' for swallowing them whole.

Seed eaters

The seed-eating finches use the bill like a hand with three fingers: the two rigid horny mandibles and the tongue as a very nimble middle finger. To de-husk a seed, a finch squeezes the seed between the mandibles to split the husk; then rotates it with the tongue so that the lower half of the bill peels off the husk to reveal the kernel. While juggling one seed, the tongue can hold several more, so the finch can gather a beakful and retire to a safe perch to eat them.

The species of finches have different bill sizes for dealing with different kinds of seeds. The short bill of a Bullfinch is used for peeling seeds and the massive bill of a Hawfinch for cracking stones of cherries and sloes. The Goldfinch and Siskin have fine, pointed bills. They prise open seedheads, often hanging upside down to do so. Dandelions and thistles are preferred by the Goldfinch, while the Siskin specializes on the cones of alders.

ABOVE: The Blackbird has a general purpose bill for a varied diet but the hooked tip and wide base are specializations for eating fruit.

A varied diet

Despite these nice links between the shape of the bill and the bird's preferred food, they are only a rough guide to its feeding habits. Each finch eats a variety of seeds and they also turn to catching insects, especially for feeding their nestlings. Even the Crossbill, which has the most specialized bill of all our birds, adds a little variety to its menu. The crossed tips of its bill are used for extracting seeds from cones but this strange bill can be used to attack apples to get at the pips or to remove flakes of bark and expose insects hiding underneath.

RIGHT: Like other finches, the Hawfinch is a seed-eater but its bill is strong enough for cracking cherry stones.

Feeding behaviour

A flock of Greenfinches and Chaffinches gathering under a seed feeder is a delightful sight, with the greens and yellows of the former mixing with the pinks and blues of the latter, but watch for the difference in the way they are feeding.

Both species glean seeds spilt on the ground but it is mainly Greenfinches that land on the feeder overhead and pull out seeds. This is an extension of the way that the two finches deal with their natural food. The main winter food of Chaffinches is seeds that have been shed onto the ground and they are picked up with rapid pecking movements. Greenfinches are more versatile and are at home on the feeder because they naturally feed on seeds while they are still on the parent plant. Chaffinches have difficulty grasping and pulling with their bills so they cannot tug seeds out of seedheads. Although this puts them at a disadvantage because they have to wait for seeds to drop, they are better equipped to pick up seeds after they have fallen and become embedded in the ground.

ABOVE: *The Brambling is a regular winter visitor to British gardens. It can be seen feeding on the ground with its close relative, the Chaffinch.*

Footwork

Some birds use their feet to steady or manipulate their food; others never use their feet. The finches illustrate the difference. I once returned home from holiday to find that a scattering of dandelion clocks on the unmown lawn was attracting small flocks of Goldfinches. They usually landed on the stems of the dandelions and weighed them down to the ground but, occasionally, a finch would reach up and pull the stem down with one foot. Other members of the finch family like Siskins and Redpolls also regularly use their feet when feeding. For instance, they pull up hanging catkins with the bill and clamp them to the twig while feeding on them. Bird fanciers used to exploit this co-ordination of foot and bill to teach captive Goldfinches to draw up a string with food tied to the end. Greenfinches and Linnets sometimes steady their food under a foot; but Chaffinches and Bramblings never use their feet to hold their food.

CHEWING WITH GRIT

With no teeth, birds cannot chew their food and have to swallow large chunks whole. I have seen a diminutive Blackcap swallow a cherry, although it had some difficulty getting it down. Finches deftly dehusk seeds between the tongue and edge of the bill but the kernel is swallowed intact. To overcome the lack of teeth for masticating their food, plant-eating birds swallow grit or fragments of snail shell which lodge in the gizzard, a muscular part of the stomach. As the gizzard contracts, the grit grinds tough vegetable tissues into an easily digestible pulp. The grit passes through with the food so it needs constant topping up. This may explain why Woodpigeons, Starlings, House Sparrows and other birds sometimes appear to be pecking at bare ground.

Feeding times

I have a bird feeder on the lawn in front of my study and, being an early riser, I once kept a record of the time that birds started to feed. To my surprise they did not come at first light but an hour or so later.

I had expected that, after the long hours of darkness, birds would need to break their fast as soon as it is light enough to see. The probable reason for the delay is that carrying a full 'fuel load', either as food stored in the crop or as body fat, makes flying more strenuous. Birds need to keep their weight down. So the early bird is not catching the worm but carrying out other duties, perhaps preening or beating the bounds of its territory. One explanation of the dawn chorus (p67) is that male birds are using 'first light' as a time to inform their neighbours of their presence by a concerted session of singing. Thereafter, feeding takes place in bouts through the day. Great Tits feed mainly during the morning and evening in winter but they have a 'working day' of 14–18 hours – almost as long as the day – when there is a brood of nestlings to feed.

*ABOVE: A Spotted Flycatcher works hard by flying from perch to catch each insect. **BELOW:** A windfall apple makes an easy meal for these Blue Tits who peck into the flesh to get the nutritious seeds.*

Feeding frenzy

If there is a shortage of food in winter small birds may be obliged to spend all the available daylight searching for enough to eat to see them through the long, cold nights. Much also depends on the type of food available. Goldfinches have to spend 8–13 hours a day feeding on the tiny, awkwardly packaged seeds of sowthistle or groundsel. They understandably prefer seeds that are more nutritious and easier to pluck. For example, they need only four hours' feeding per day on dandelions, and one hour on teasel or burdock. When feeding is easy, cold weather and short days are not a problem and birds have time for other activities and even to 'loaf', doing nothing in particular.

Efficient feeding

To keep well fed and leave plenty of time for other activities birds must feed efficiently. Their food often occurs in patches – a birdfeeder, a berry-laden bush, an aphid-infested plot of peas. So a bout of feeding starts with a search for a new source or a return to a known supply. During the course

ABOVE: *A Waxwing demonstrates gymnastic skills in order to reach a cluster of ripe berries.*

of the day, the birds will visit a number of food sources and stop at each one for only a short time. This can be seen from watching a birdfeeder. Despite the almost unlimited food, flocks of tits and finches drop in, feed intensively for a while, then move on. I have a feeder filled with nyjer seed which attracts Goldfinches. The resident pair visits it for only a few minutes each day, so it provides no more than a small part of their daily diet. Bird-ringing has shown that there is a steady passage of birds through a garden. A hundred Blue Tits can visit in one day, each one stopping off as it works its way around a feeding range extending over several acres.

Getting it right

Birds try to choose food that gives a good return of energy for time spent gathering it. The basic rule is: not too big, not too small, but just right. For instance, Spotted Flycatchers prefer to hawk for medium-sized insects, such as hoverflies and blue-bottles. Each one is a good package of food and easy to handle. Small insects such as aphids are not worth the effort of catching – unless they are very abundant or there is nothing better available. Large insects like dragonflies are big packages but they are less economical because they take so long to hunt, dispatch and dismember.

The basic rule can be broken in special circumstances. We once had a pair of Spotted Flycatchers nesting in a corner of the house where there was a brick missing. Their characteristic 'flycatching' provided us with plenty of entertainment as they flew out from a perch, snapped up an insect and returned to the perch. Butterflies did not feature in the diet, although they were abundant, because their erratic flight made them difficult to catch. This pair of flycatchers was unusual, however, because it reared a second brood, by which time the buddleias were in bloom and attracting a batch of butterflies. They were newly-emerged from the chrysalis so their wings had not fully hardened and their flight was still weak. They fell an easy prey to the flycatchers.

The rule of choosing prey that gives a good return for effort applies to thrushes eating snails. Song Thrushes are famous for smashing open snail shells on an anvil, usually a stone or a paving slab, which becomes surrounded with a litter of broken shells. Extracting a snail from its shell is a time-consuming process and thrushes prefer worms and caterpillars, which are easier to deal with. They concentrate on snails only when other prey is hard to find, as happens when spells of drought or frost drive worms deep into the soil. This answers the question of why snails bother to live in a shell when it does not guarantee their safety. An impregnable shell would be too heavy to carry, but light, practical armour acts as a significant deterrent and improves a snail's chances of survival.

ABOVE: A Song Thrush works hard at its anvil to extract a snail from its shell. *BELOW:* Black-headed Gulls and Starlings take to the sky to make the most of a swarm of flying ants.

Pest control

Birds are sometimes described as 'gardener's friends' because they eat pests. Blue Tits pick aphids off roses and Song Thrushes seek out snails sheltering among delicate plants. We are delighted to see them at work but it is by no means clear that they are getting rid of pests, any more than Magpies or Sparrowhawks are reducing the numbers of garden birds (pages 134-135). In Chapter 1 I described how H J Massingham was pleased to see his Blue Tits catching thousands of harmful weevils and sawflies in the course of a day. If these insects were very abundant, the tits would not have had any significant effect on their numbers. An animal's population is only reduced if it is being culled faster than it can grow.

Aphids, as gardeners know, breed incredibly rapidly, and quickly build up to pest proportions. A pair of Blue Tits, even when aided by other birds, would not have a significant effect on a plague of aphids in its territory. However, some research does show that birds can deplete the numbers of caterpillar pests in orchards sufficiently for a significant increase in the crop. They are not the complete answer to pest control but their efforts may mean that fruit-growers can reduce the amount of pesticides sprayed on the trees.

Learning new foods

If you put up a new birdfeeder, the first birds may arrive on it before you have got back to the house but it generally takes time for the neighbourhood to learn about this new fast-food outlet.

I have often been puzzled about the way birds discover a new kind of food, especially when the food is not a natural part of their diet. Some species are more adaptable and inventive than others. House Sparrows, Starlings and Blue Tits are quick to find anything new on offer but it took several years for a Great Spotted Woodpecker in my garden to realize that the sunflower seeds hanging next to the peanuts were also edible. Sometimes a bird learns

BELOW: A Great Spotted Woodpecker carries a nut to a 'vice' – a crevice where the bird will lodge the nut so that it can be hammered.

by watching another and following its example but someone has to make the initial discovery. It may be by trial-and-error as the birds search the neighbourhood and test possible new sources of food.

At one time Long-tailed Tits were rarely seen at birdfeeders. It was thought that this was because, compared with our other tits, they rarely eat seeds. This makes them vulnerable in winter when snow and ice make insects hard to find. However, in recent years, Long-tailed Tits have discovered peanuts and, from being rare visitors, flocks are now a familiar sight on feeders, even in late summer when insects are still abundant.

Eating anything edible

With birds searching every nook and cranny, nothing clearly edible can be overlooked for long. This was obvious some years ago when I started to put out food for the birds in preparation for the winter. Apart from the scraps of potato, bread, meat and other leftovers, I hung out a titbell, an inverted earthenware bowl in which molten fat is poured and left to set. At first it was ignored while the birds fed on the scraps. Then once in a while a Blue Tit or Great Tit paid a short visit. A couple of weeks later, there was a stream of tits queuing to feed. Eventually, House Sparrows found how to get at the fat, although they never became so adept at landing upside down under the bell as the tits. What puzzled me is how tits first learned that the hard, smooth surface of the fat under a titbell is edible. It is unlike any other food, animal or vegetable, that tits normally eat.

The same holds for a variety of foods that birds learn to take. When there are long, cold spells in late spring, the birds that depend on insects may be hard-pressed to find enough food. Nectar from the flowers of currants and gooseberries and willow catkins is one alternative source of energy and Blue Tits sometimes take significant quantities. However,

it is not always easy to see what birds are doing at flowers and to be sure whether they are sipping nectar, collecting pollen or even feeding on insects inside the flowers. An obvious question is whether birds help to pollinate the plants. Blue Tits have been seen with their faces and breast feathers dusted with pollen from male catkins, which could be transferred to female catkins.

The opportunism of Blackcaps

Through the Nature Note column that I used to write for the *Daily Telegraph*, I discovered that Blackcaps eat stamens from varieties of mahonia that flower early in the year. Someone wrote to say that they had seen a Blackcap delicately probing mahonia flowers and removing bunches of stamens so carefully that the rest of the flower was unharmed and clusters of berries eventually formed.

Eventually I received 26 letters recounting similar observations. There is no mention of this habit in the scientific literature on warblers so I wonder if this is a new trick that has been learned by the Blackcaps that now winter here instead of flying to Africa. It would not be surprising because the Blackcap's survival through our winter seems to depend at least in part on its opportunism. Blackcaps will turn their beaks to anything, from bread, porridge and suet on the birdtable to peanuts in feeders and the nectar of winter jasmine. So it is not surprising to find a few enterprising individuals making a meal of the protein-rich pollen in the stamens of early flowers.

Finding and storing food

How does a bird find its food? When it sets out to feed it must have some idea of what it is looking for, and when and where to find it

It is believed that animals have a 'search image' – a sort of mental picture – of what they are looking for. It may be like our 'mind's eye'. We find it easier to find something we have lost if we have a good idea of what it looks like. From observations on baby birds, it is known that the search image is partly instinctive; in other words birds are hatched with the knowledge of what to eat, but they also learn by experience. A fledgling Chaffinch, for example, starts by pecking at anything that contrasts with the ground and picks up objects and tests them in its bill. It gradually learns to discriminate between what is edible and what is not and finally it develops a preference for the largest seeds that it can comfortably handle.

Sharp-eyed birds

Sight is the most important sense for finding food. It used to be said that a Blackbird standing with its head cocked, before suddenly thrusting its bill into the ground and seizing an earthworm, was listening for the faint rustling of the worm's tiny bristles against the sides of its burrow. It is now known that the Blackbird is more likely to be looking for the protruding tip of the worm's body.

ABOVE: *Nuthatches store nuts and seeds by hammering them into crevices and sometimes covering them with moss or bark.*

RIGHT: *Although it appears to be hunting by ear, a Mistle Thrush looks for the tip of a worm protruding from the soil.*

Birds' colour vision is as good as, if not better than, ours although the acuity of vision is probably not so good. However, birds appear to be exceptionally sharp-eyed because they are continually looking about them and noticing small details that are a matter of life and death. For example, a Swallow was seen to drop a honeybee, dead, to the ground. It proved to be a stingless male or drone. Stinging worker bees and wasps are well protected from attacks by birds. Once stung, the bird learns to leave them well alone although flycatchers rub the insects' abdomens against a perch to squeeze out the venom before eating them. Swallows, which don't retire to a perch to eat, cannot do this. So it is not surprising that they have been recorded catching only drone honeybees. A Swallow must make a quick, and very accurate, decision as to whether its target is a stinging worker or the slightly larger harmless drone.

ABOVE: Jays hide acorns and peanuts in the ground and remember the precise locations so they can retrieve them later.

Storing food

Crows and their relatives are wary birds, with justification because of their long history of persecution. They rush into the garden, gobble as much as possible from the birdtable and make a quick getaway. When they leave, there is a visible bulge at the base of the bill where food is crammed in a pouch. The pouch is important for carrying food to the nestlings, but it is also useful for birds that find more food than they can cope with on the spot. The surplus is carried away and buried. The bird pushes its beak into the ground or other suitable nook or cranny and inserts the food. It is then covered over with soil or dead leaves and abandoned, but not forgotten. It is dug up and moved if another bird shows an interest and

it will eventually be recovered and consumed. Jays are specialist hoarders and their distribution is linked to the occurrence of oak trees. During the autumn they make the most of acorn crops. Any acorns not eaten immediately are carried away and stored under dead leaves or in a hole dug with the bill. The location is remembered so precisely that acorns can be retrieved from under several centimetres of snow. And, next summer, when any acorns that have been overlooked sprout, the Jays pull up the seedlings and feed them to their nestlings.

The Coal Tits' cache

Coal Tits are not the most frequent visitors to bird-feeders but they are very industrious and have a disproportionate effect on the supply. While other birds peck each morsel to break it up and swallow it, the Coal Tit runs a shuttle service, grabbing a seed and flying off with it, then returning a minute later for another. If you find the other end of the tit's flight path, you will see where it is hiding its booty. Food is hidden in a variety of places. It is pushed down crevices in bark or cracks in fences, stuffed inside bunches of conifer needles, or buried in the ground. Sometimes the cache is covered with soil, moss or a flake of bark. If there is not a birdfeeder handy, seeds are collected from plants, especially from pine or spruce cones. Animals such as caterpillars and aphids are also hoarded. However, unlike crows and jays, there is little evidence for Coal Tits returning to their hoards except by accident. It is known that Marsh Tits remember the exact location of each hiding place for a day or two, but the Coal Tit has not been the subject of detailed study.

ABOVE: Jays help the spread of oak by carrying acorns away from the trees. Those that are not eaten later will sprout.

LEFT: Coal Tits industriously hide seeds and other food but they are unlikely to retrieve them.

CHANGE OF DIET

Diets vary through the year as different foods come into season. Insect-eating birds like tits and warblers feed extensively on fruit and seeds in winter and the specialist seed-eating finches turn to catching insects especially in summer when they have nestlings to feed. Greenfinches, Bullfinches and Goldfinches feed their broods on a mixture of seeds and insects while themselves remaining strict seedeaters. It seems that insects provide the protein and minerals needed for the growing bodies of the nestlings, but they may also be easier to digest. However, some young finches, including those of Siskins and Linnets, are reared on a pure seed diet. Great Tits bring progressively larger insects to their nestlings as they grow older but bring more spiders to young nestlings, perhaps because they are rich in a particular amino acid needed for growing feathers.

I was amazed to read how Great Tits change the shape of their bills through the year in order to adapt to changes in diet. Many birds wipe their bills on perches to keep them clean. The stickier the food, the more they wipe, but Great Tits strop their bills to change their shape. In summer the bill is slender and used as tweezers to pick up small insects. In winter it becomes shorter and deeper, and better for handling the winter diet of seeds. The change in shape is assisted through bill-wiping which hones the edges of the bill. By adjusting the amount of wiping, the Great Tit quickly matches the shape of its bill to a new diet.

ABOVE: Bullfinches feed on seeds and buds all year but give insects to their nestlings. BELOW: Starlings gather on a lawn to hunt for leatherjackets hiding in the grass.

Birds and fruit

There are two completely opposing views on the subject of birds and fruit. While one gardener is delighted to see Redwings and Fieldfares descend on cotoneasters and strip them bare of berries, another is outraged when a crop of cherries is ravaged by Starlings.

A neighbour once remarked how silly it is to put out food for the birds in winter and then spend the summer trying to keep them off the peas and soft fruit. Such an ambivalent attitude – tempering a concern for the welfare of birds with annoyance at their depredations – is relatively new in the history of human relations with birds.

My father recalls how, in the early years of the last century, boys in his village threw stones at Blackbirds as a matter of course. This was not out of

BELOW: A diminutive Song Thrush has to use its weight to pluck berries from bushes.

devilment but knowing that Blackbirds stole the fruit that was part of their families' livelihood. Even that tiny innocent, the Blue Tit, was once persecuted for theft. Although its tiny, needle bill is designed for picking caterpillars and aphids off leaves, the Blue Tit sometimes attacks peas and beans in their pods and pecks holes in apples and pears.

A symbiotic relationship

It may not seem so at first sight but there is often a mutual relationship between the eaters and the eaten. From the plants' point of view, their fruit is designed to be eaten. The nutritious flesh and eye-catching colouring help to ensure that seeds are distributed far and wide. Birds eat the fruit, digest the flesh and either regurgitate or pass the seeds at some distance from the parent plant. You can sometimes see a Blackbird or Robin 'heaving' and suddenly bringing up a small mass of seeds.

The most familiar example is that of mistletoe and the Mistle Thrush. The sticky mistletoe fruit adheres to a branch, either when wiped off the thrush's bill or after it has passed, undigested, through the bird. The seed sprouts the 'roots' that connect to the host tree and enable the mistletoe to live as a parasite by tapping its sap.

Some birds, such as finches and tits, cheat plants by eating seeds and digesting them, so the plant's attempt at reproduction comes to nothing. Tits also thwart plants by pecking the flesh of large fruits, such as apples, and ignoring the seeds so they are not dispersed.

Fruit all year long

There is a succession of fruiting from early-ripening cherries, through blackberries, haws and yews followed by hips, holly and privet. One of the last garden plants to ripen is ivy which fruits in January or later. Some fruits, especially holly and ivy and, to a lesser extent, haws, yew and crab apple, stay on the tree and remain in an edible condition until the following summer. The seasonal ripening of fruit gives birds a bonanza when they need high-energy food to keep warm. We are all familiar with the way that holly, cotoneaster and rowan are stripped bare within a few days of birds discovering them. In the severest winter weather, fruit exposed at the tops of branches thaws

ABOVE: A Blackbird surrounded by food. Windfalls provide easily obtained meals through autumn and into winter.

while the ground is still covered with snow or locked up by frost. Fruit gives the birds an easily-procured meal and it is particularly annoying if birds find the holly berries that are being saved for Christmas decorations!

According to folklore, a good crop of fruit or berries is a sign of a hard winter ahead, although it is a mystery how the trees foretell the future. What is certain is that a good crop is a sign that, if the winter proves to be bad, many birds will find enough food to support them through hard times. So whether you see birds as a cheerful attraction in winter or a plague on ripening fruits in summer is a matter of your own priorities.

Leaf eaters

Woodpigeons are unusual because not many birds eat leaves. Compared with the variety of leaf-eating

ABOVE: Woodpigeons are one of the few birds that regularly eat leaves. They are often seen eating fresh hawthorn leaves in spring. BELOW: It is rare for a Tawny Owl to go fishing and it probably happens when other food is scarce.

mammals, from voles to elephants, only a few birds have a diet of leaves. These include geese, swans, ducks, grouse, partridges, Moorhens and Coots. Leaves are difficult to digest and not very nutritious so birds prefer soft, young leaves which are rich in protein. You can test the palatability of leaves for yourself. Fresh leaves of hawthorn are edible, and used to be called bread-and-cheese by country children, but old leaves are unpleasantly difficult to chew. As foliage matures, it becomes tougher and the tissues accumulate chemicals such as tannin which further reduce its digestibility. These chemicals provide defence against caterpillars and other insects. When a leaf is chewed, the chemicals change the leaf proteins so they become hard to digest. It is the same process as tanning leather. Caterpillars have such difficulty digesting tanned proteins that they may fail to develop into butterflies or moths. And nestling tits fed on these caterpillars grow slowly and may not fledge.

Drinking

Gideon chose his select band of Israelites to attack the host of Midian with a drinking test. Those that drank by lifting water in cupped hands were selected. Those that put their faces down to the water and lapped like dogs were rejected.

You will see that the majority of birds dip their bills in the water and then lift their heads to swallow. Woodpigeons and Collared Doves, however, are the equivalent of Gideon's lappers and he would have rejected them.

Gideon's test was designed to select soldiers who remained alert and on guard even when drinking. Birds that take a sip and then look around as they swallow are less likely to be surprised by a cat or hawk than a bird that keeps its head down. The question is, then, why does the pigeon family (and a few other families) 'lap' or, more correctly, suck up water. Some experts believe it enables these birds to drink quickly and spend as little time as possible in a potentially dangerous situation. Moreover, pigeons and doves have bulging eyes that give them an almost all-round vision and make it extra difficult to catch them napping. Swallows and Swifts avoid the danger of being caught drinking by sipping water as they swoop at speed across the surface of ponds.

Water from food

Although frequently seen drinking at garden ponds and birdbaths, birds do not need to replenish the water in their bodies so much as mammals. Their metabolism produces uric acid as a waste product which is voided in the droppings as a semi-solid white fluid. This is the 'whitewash' that accumulates under bird perches and can become a problem under pigeon and Starling roosts. Mammals produce urea which must be flushed from the body with a copious flow of water. Neither do birds lose water by sweating but they do lose a certain amount in their breath.

Small birds may need to drink several times a day in hot weather but in cool conditions some species can survive without drinking. The need to drink is reduced when the food is succulent, such as juicy fruits, worms and insects. Nestlings, for instance, do not receive water from their parents and must rely on water in their food.

Frost locks up the water supply but birds may need more water in winter when they are more likely to be feeding on dry seeds. We are encouraged to keep our birdbaths free of ice but when there is no free water, birds resort to eating snow. They

ABOVE: *Birds naturally conserve their body water but there are times when they need to make good their losses by drinking.*

even use the same action as drinking: pecking at the snow then raising the bill and making swallowing motions. This is literally a chilling experience because the bird has to use its precious body heat first to melt the snow and then raise the temperature of the water to body heat. This could be a life-or-death matter for a starving bird that is already having difficulty keeping warm.

Songs and signals

'Animals do not possess a language in the true sense of the word….every individual has a certain number of innate movements and sounds for expressing feelings. It also has innate ways of reacting to these signals.'

KONRAD LORENZ KING SOLOMON'S RING

Konrad Lorenz won the Nobel Prize for his pioneering scientific studies on animal behaviour and his delightful book *King Solomon's Ring* was an inspiration when I was a boy. It helped to teach me the importance of curiosity in observing animal life and also that the study of animals does not require travel to exotic locations. The research on animal communication that Lorenz describes took place within the confines of his house and garden. The title of the book is derived from myth that King Solomon could talk to animals. While we may not able be able to converse across the species divide, the work of Lorenz and his successors has given us a good idea what birds are communicating to each other by voice and gesture.

ABOVE: *Mistle Thrushes used to be called 'stormcocks' because they sang from exposed perches while they were lashed by gales.*

LEFT: *The Blackbird's song is one of the garden favourites, especially when it is heard on a fine evening.*

Song

Three hundred years ago, Joseph Addison, poet and statesman, wrote 'I value my garden for being full of blackbirds than of cherries, and very frankly give them fruit for their songs.'

The song of a Blackbird is one of the best-loved sounds in the garden and a stroll in the garden of a summer's evening is made perfect by the late chorus of the neighbourhood's Blackbirds. Birdsong is one of the glories of the natural world that anyone can enjoy simply by going into the garden in the early morning or evening. This is what poets write about, but Robert Browning's 'first fine careless rapture' of the Song Thrush is a personal, subjective assessment. Ornithologists are concerned with the prosaic matter of function and how the Song Thrush, and any other bird, interprets another's outpourings. Not for them the romantic notion of a bird singing for its own or its mate's pleasure. The birdsong that was once the province of poets is now subjected to rigorous and objective scientific study. This does not reduce the pleasure of listening to birdsong; rather, science enriches our appreciation.

Defining birdsong

The first step to understanding birdsong is to define it. A simple definition that fits most garden birds is a 'form of vocal communication which is musical to our ears'. This omits from the definition voices that are anything but musical. Gilbert White commented that 'Rooks in the breeding season, attempt sometimes, in the gaiety of their hearts, to sing, but with no great success'. Blackbirds, Song Thrushes and Blackcaps have richly varied songs while Chiffchaffs and Cuckoos do little more than utter the two notes that give them their names. The male Chaffinch sometimes utter a hoarse 'tweet' which is repeated metronomically every second for minutes on end. Then it flies to another tree and starts again. These monotonous notes appear to have the exactly same function as the more familiar jangling song. So it is a puzzle why the Chaffinch should bother with the complex series of notes that makes up its normal song when the monotonous monosyllables do the same job. Perhaps the former conveys more information, in the same way as a greeting is so much more informative if it includes a remark about the weather rather than just a curt 'Hello!'

BELOW: The apparently simple song of the Great Tit proves to be variable and each individual has a repertoire of variations.

Evidence for this comes from the Great Tit. Its triple-phrase 'teacher - teacher - teacher', like a squeaking bicycle pump, is much more variable than this simple transliteration suggests. Each bird has several phrases which differ in their number of notes and in changes in pitch and amplitude. They are sufficiently distinct for females to recognize the songs of their mates. There is also some evidence, in Great Tits and other species, that variety in the song of an individual is more effective in driving off rivals than a set of unchanging notes. In an experiment, recordings of Great Tit songs were played through loudspeakers in a wood. If a patch of wood was 'defended' by a simple repeated song, it was quickly occupied by male Great Tits. But if the song was varied, they stayed away.

The Beau Geste effect

So it seems that by changing its song a male Great Tit fools other tits looking for somewhere to settle into thinking that they are in an area already fully settled. This has been called the Beau Geste effect, after the ruse in the novel of that name, when hard-pressed legionnaires in a desert fort propped up the bodies of their fallen comrades behind the battlements and ran to and fro firing their rifles to give the impression that there were a large number of defenders. This cannot be quite the whole explanation for the variety in birdsong since Mistle Thrushes, for instance, sing their repetitive phrases from regular, exposed, perches so newcomers to a neighbourhood will quickly learn the whereabouts of the inhabitants.

MEMORABLE SONGS

Remembering a song is always made easier if it can be set down in words as an aid to memory. Not everyone hears a song in the same way, as is shown by the different names for a Cuckoo in foreign languages, but a transcription can be matched with the real song.

Here are some transcriptions of garden bird songs by different authors:

Chaffinch: *Sweet, sweet, sweet, pretty lovey, come-and-meet-me-here.*

Greenfinch: *Clip, clop, clump, sneeze.*

Great tit: *Teacher-teacher-teacher.* Once called the 'saw-sharpener' because its two see-sawing notes sound like a file being drawn to and fro on a metal edge.

The Linnet's song is a medley of musical notes, starting with a trill.

Woodpigeon: *Take two coos, Taffy or My toes hurt, Betty.*

Yellowhammer: *Little bit of bread and no cheeeese.*

Song thrush: *Knee-deep, knee-deep, knee-deep; cherry-do, cherry-do, cherry-do, cherry-do; white-hat, white-hat; pretty-Joey, pretty-Joey, pretty-Joey.*

Nightingale: *Wheet, wheet, kurr, k-u-u-r-r-r. Sweet, sweet, sweet, sweet, Jug, jug, jug, jug, jug, Swot, swot, swot, swotty (sweet, sweet is plaintive and jug, jug is quick like a dog barking).*

Goldfinch: *Sippat-sippat-slam-slam-slam-siwiddy.*

Linnet: *Hepe, hepe, hepe, hepe, ollaky, tollaky, quakey, wheet, lug, orcher, wheet.*

What are they singing about?

Charles Darwin considered that song was 'for the charming of females'. Perhaps he was unduly influenced by the beauty of birdsong, and during the 1920s, when the significance of territory in the life of male birds became to be appreciated, opinion changed to the concept of song as a 'message of hate' directed at other males.

Nowadays, it is accepted that song has two functions: attracting females, repelling males, or both. It is a beacon that allows other birds to pinpoint the position and identity of the singer.

Establishing territories

Most birds sing most intensely at the start of the breeding season, continue throughout the nesting period and stop when their offspring are becoming independent. A few species, such as the Sedge Warbler, stop singing as soon as they have paired. When males start to sing, they are intent on establishing their territories before they pair and eventually settle into nesting. Their attention is focussed on other males and they sing most passionately when other males intrude into their territories. The Wren has a powerful voice for a small bird. When two are arguing over their territory boundary, they sing antiphonally, a chorus from one followed by an outburst from the other, with such vehemence they sound as if they are about to explode. Gradually, as the birds settle into their territories and establish the extent of their claims, the volume of song is reduced. Each bird now knows its place and recognizes its neighbours so there is less need to spend time proclaiming the territory boundaries. Neighbours can even intrude into the territory without causing an angry reaction. A stranger, however, will still provoke an outburst of song.

Quality of song

The second function of birdsong is to attract, and maintain, a mate. It is easy to overlook because a dispute between two males is obvious compared with a female quietly listening and watching. Female birds are, however, not passive and simply acquiescing to the blandishments of the first male they meet. As will be described in Chapter 8, they try to mate with the best male available. One way of selecting a good male is by the quality of his song. The males of some species develop longer and more varied songs as they get older. A bird that survives to a good age is likely to be a superior individual in good condition and therefore a good father. So a good song is the sign of a high quality male. In one study, Dunnocks that were supplied with supplementary food started singing earlier in the year, sang more frequently and had more success with breeding. One conclusion to be drawn from this is that, like Joseph Addison, you should encourage birds with plenty of food if you want your garden filled with birdsong.

BELOW: The Collared Dove's monotonous coo can be mistaken for the Cuckoo's two syllables.

ABOVE: *The Robin is unusual because the female sings to defend her own territory in winter.*

Singing females

Back in the 1920s, the naturalist W. H. Hudson knew that the female Robin sings as well as the male. He argued that because every Robin has a territory in autumn, both males and females must hold territories. And, as all resident Robins sing, both male and female must be singing. It is now known that female Robins hold their own territories in winter and sing to defend them, although they are not so vocal and their repertoire is not so elaborate as that of the males. Females continue to sing in spring when they start to pair up and fall quiet when they become absorbed with nesting. Thereafter, the male continues to sing in defence of their shared territory.

The female Dunnock may sing during the breeding season to solicit the company of the male. As he may mate with two or three females, she does not get his undivided attention and an element of competition is introduced. As with the Robin, singing female Dunnocks go unnoticed because the sexes are alike. The female Starling, on the other hand, sings to keep other females away from her mate while the female Great Tit sings to contact her mate.

DRUMBEATS

Drumming is a form of song and is used as a signal by Great Spotted Woodpeckers of both sexes. It is most often heard around the nest-site during the breeding season, from January to June and again in September. Its function is to help to keep members of a pair in touch with each other and the male also uses it as a kind of song to court prospective mates and deter rival males.

A Great Spotted Woodpecker treated me to an exhibition of drumming. I watched it climbing the trunk of a poplar tree and pause at intervals to deliver a series of blows so rapid that they run together to make a sound that resembles a sharp, nasal snore or the creak of an extremely stiff door. Drumming can sometimes be heard over half a mile so the blows must be delivered with great force as well as speed. The woodpecker saves itself from brain damage by the special structure of the head and neck which cushions the force of the blows.

The song period

One Christmas Day, we heard an extraordinary chorus of Mistle and Song Thrushes. Their repeated phrases sounded louder than usual, perhaps because this is the one day of the year when the noises of civilization are least obtrusive.

That birds should be singing so strongly around the winter solstice, and some weeks before we expect the hardest winter weather, is an indication of how vital it is for a bird to maintain its claim to a territory whenever possible. Blackbirds normally start to sing in February but they may start as early as December and I have heard one short snatch of song as early as October. Town Blackbirds start singing before their country cousins, perhaps because of the extra warmth and shelter provided by the urban environment. The first Blackbird song is heard mainly when the light is just beginning to fade on mild, still, damp afternoons. Only later do Blackbirds join the dawn chorus. The first singers are predominantly young cocks setting up a territory for the first time. The best time for enjoying Blackbird song comes later, from March to June, when the older cocks join in while their hens are on eggs. At this point the Blackbird dominates the dawn chorus. Once the cocks are helping to feed nestlings, the time spent singing diminishes and by July they have fallen silent.

ABOVE: The clear, musical notes of the Song Thrush may be heard for much of the year.

The quiet period

Most garden residents have a much longer song period than the Blackbird. After the full ebullience of their song during the main courtship period in spring, they become quieter while absorbed with the care of their families, and fall silent when the young have flown. This is the silent time for bird song and it is a sign that summer is drawing to a close. Silence falls because the breeding season has finished. A renewed bout of singing starts in September and is sometimes strong enough to give the impression of a return to spring. Robin, Wren and Great Tit are much in evidence but others, such as Blackbird and the finches, remain mostly silent. Pigeons and doves continue to be vocal during this period because they have a prolonged breeding season.

Late summer singing

If it takes more than one Swallow to make the summer, Chiffchaffs and Willow Warblers make the Indian summer. Every year around late August and September I expect to hear snatches of song from one or other of these warblers when they stop for a day or two in the garden. In March these songs proclaim the warblers' return from warmer countries and the imminent arrival of spring. Now, they are like a treat at the end of the school holidays: an immediate pleasure but also a reminder of less congenial times ahead.

The sound of winter

The Starling is a good winter songster when the thrushes have gone quiet. Small flocks gather at the tops of their trees and give a performance as the sun sets. One note, in particular, identifies the Starling, and that is a clear whistle and descending note. But there are all sorts of other sounds – sometimes just a cacophony of whistles, squeaks and clicks, but also more musical and even delightful notes. Some of the whistles make you think that a Blackbird has started to sing unseasonably early until chattering and chuckling show that this was just part of the Starling's repertoire.

LEFT: The Starling is underrated as a songster and enlivens winter afternoons, although its song is sometimes swamped by chattering and squeaking.

IDENTIFYING SONGS

Every spring I have to remind myself of the identity of some songs. This is often frustrating when the singer is only a blur moving through the foliage. CDs and portable players have revolutionised the process of learning birdsong but there is still a confusing variety of songs for novices to sort through. Fifty years ago Len Howard, a musician who chronicled the birds of her Sussex garden (*Birds as Individuals*, Collins 1952) described four main types of song, which helps to narrow the possibilities when tracing an unknown song.

ABOVE: The tiny Wren has a powerful voice and sings a set but elaborate song.

Chiffchaff, Blue Tit, Great Tit, Coal Tit, Swallow, Collared Dove, Stock Dove, Woodpigeon, Feral Pigeon, Bullfinch, House Sparrow.

Type 2. Birds whose songs conform to a set but much more elaborate rhythm: Chaffinch, Yellowhammer, Wren, Goldcrest, Dunnock, Greenfinch, Mistle Thrush, Nuthatch, Redstart.

Type 3. Birds whose rhythms are free and varied, the accented beat changing: Blackcap, Linnet, Goldfinch, Starling.

Type 4. Birds whose song consists of many distinctly different phrases in different rhythms, sung with a moment's pause between each phrase:Blackbird, Song Thrush, Robin, Nightingale.

Type 1. Birds that have only one song in a set rhythm of the simplest possible form: Cuckoo,

Learning the tune

Early experiments showed that birds reared in isolation do not develop a complete song. They produce only the instinctive core and need to hear other birds to develop the full repertoire.

Some young birds start to sing not long after they have left the nest. Some have even been recorded as singing while still nestlings. The full song only develops during the first year of life and learning is aided by hearing the songs of neighbours.

The Chaffinch is one species that takes time to find its full voice. At first it utters the tinkling notes descending the scale and omits the final flourish. So you hear 'Sweet, sweet, sweet, pretty lovey' but no 'come-and-meet-me-here'. Because the song is stereotyped and repetitive, it is easy to tell if it is incomplete. The song develops through a combination of instinct and learning. A young Chaffinch starts to sing shortly after leaving the nest but produces no more than a quiet, rambling selection of notes. It is only after its first winter that it starts to compose its full song. At first there is still no more than a simple series of notes. This is the instinctive framework. Then it adds phrases learned from its father's song, which it heard while still in the nest, and next it incorporates parts of its neighbours' songs.

One result of this imitation is that Chaffinches have regional dialects. All the Chaffinches in one place sound alike but, the farther you move away across country, the more different the local Chaffinches will sound from your home ones. The pattern of the songs also changes slowly with time

ABOVE: *The Greenfinch's distinctive song is usually given from a perch but also during a special song-flight characterized by slow, deep wingbeats.*

because mistakes made while learning are passed to the next generation.

Subsong

Listen carefully next time you are in the garden, especially in spring but also in autumn and winter, and you may hear a Blackbird, or perhaps a Robin or Chaffinch, singing very quietly. It sounds rather like someone singing to themselves as they go about their business. One ornithologist described a young Blackbird as sounding like 'a shy amateur testing his talents'. This is subsong. It is typically a rambling, formless collection of varied notes with a wider range of pitch than normal song. There is a ventriloquial quality which is heightened by the bird singing with its bill closed or even blocked with food, while the performance often takes place in the cover of a bush rather than from an exposed perch. The Blackbird's subsong is the one most often heard in the garden and it regularly incorporates alarm calls and mimicked notes of other species.

Ornithologists used to be puzzled by subsong because it had no obvious function and the birds seemed to be performing for their own benefit. Current opinion is that subsong may be a form of play, with the same functions of practising and learning skills and perfecting social behaviour. The bird listens to itself and attempts to match its

MIMICRY

As well as imitating their neighbours as part of the essential process of learning their song repertoire, some birds are mimics of other species. I once noticed the flight calls of a flock of Greenfinches sounding clearly between the noisy rattles and squeaks of a Starling singing at the top of my walnut tree. I looked up, expecting to see the flock of finches pass overhead, but the pale blue winter sky was empty. Slightly puzzled, I resumed raking the dead leaves. The calls came again, but still there were no finches in sight. Then I realized that the twittering chorus was part of the assortment of sounds coming from the Starling. This is not the first time that a Starling's skill at mimicry has caused me to doubt the evidence of my ears. Once, on a Shetland island, I heard a Curlew piping from inside a dry-stone wall. Peering unbelievingly into a deep recess, I flushed out a Starling. Another surprise came from a Tawny Owl that was not only hooting repeatedly by day but moving from tree to tree. It turned out to be a Jay hooting just like an owl.

ABOVE: Young male Blackbirds first start to sing fitfully in autumn. The full territorial song develops in late winter when the final territories are established.

vocalisation with the syllables which it had learned from other birds when it was young. Support for this idea comes from the fact that birds whose song is instinctive and not learned do not have a subsong.

The Starling's song is a shapeless medley of squeaks, rattles and whistles to which just about anything can be added, from the crying of a baby to the buzz of a chainsaw cutting through a tree – and even the sound of the tree falling. It is something of a puzzle why Starlings, and some other birds, include both the songs of other species as well as completely unbirdlike sounds into their own songs. One possibility is that mimicry is their way of improving the song repertoire to make them sound old and experienced.

LEFT: The Greenfinch is one of many garden birds that incorporates the notes of other birds into its song.

Dawn chorus

I am not sure whether the dawn chorus is a blessing or a curse for insomniacs. I have found that counting 'cuckoos' banishes sleep but the full dawn chorus is a compensation for sleeplessness.

One morning in early May, I woke an hour or so before dawn. There was not even a hint of sunrise and I was in time to pick out the different songs as they started. A Song Thrush started fitfully and took a minute to get into the full rhythm of its repeated phrases. Then a Blackbird chimed in with its fruity notes, followed by an outburst from a Wren. By then I was fully awake so I got up and went into the garden. The light was that mysterious grey that precedes dawn and Blackbirds and Robins were in full voice but I had to wait for Chaffinch, Blackcap and others, which only joined in when it was fully light.

LEFT: *The loud, raucous notes of a cock Pheasant sometimes precede more musical renditions from other species and herald the start of the dawn chorus.*

After about half an hour all fell silent again and I went back indoors, but some time later I was aware that the chorus had started once more. It used to be said that this quiet period was when the birds were 'at their prayers' but an ornithologist who made a study of the dawn chorus many years ago was nearer the mark when he suggested that this time between choruses is when the birds 'have their breakfast'. When the sun is up singing becomes intermittent. It dies away in the afternoon but there is a dusk chorus before the birds go to roost.

The dusk chorus

Less well-known than the dawn chorus, the dusk chorus starts towards nightfall. The loudest parts are the Blackbirds' 'chinks', which appear to serve the same function of advertising the vocalists' presence as their musical song. The chorus peters out as the darkness deepens and the birds settle into their roosts. The Robin is one of the last to stop and may sing so late into the night, especially under street lights , that it is mistaken for a Nightingale.

The function of the chorus

The function of the dawn chorus still puzzles naturalists, although there have been plenty of theories. A male bird has two priorities when it wakes on a spring morning: it needs to feed and it must remind its neighbours of its territorial claim. It does not start singing at once but rouses gradually by spending a few minutes stretching and yawning before flying to a song post. Breakfast might seem to be the priority and the early bird ought to be getting the worm but the search for food cannot start until it is light enough to see clearly. It only gets light enough for Great Tits to find insects after their dawn chorus has come to an end. So the birds start the day by 'beating the bounds' of their territories with song.

The order in which species start singing is not invariable. Often it is the harsh notes of a Pheasant or Carrion Crow, or the gentle cooing of a Woodpigeon that lead off. The sequence is determined by the ability to see in dim light. As a general rule, birds with larger eyes, like the thrush family, start singing first and smaller-eyed Chaffinch and Great Tit begin later, perhaps even after the early birds have stopped.

Call notes

As well as songs, birds also have a variety of simpler calls, usually of a single note, which are used in different contexts. They are given by birds of both sexes and all ages and include the begging calls of hungry nestlings, alarm calls warning of danger and contact calls for keeping birds in touch with each other.

The greatest repertoire

There are calls given in territorial contexts in addition to the song. The Chaffinch has a set of over a dozen different calls for all occasions and the Great Tit has 18 with some additional variations. It is often said that if you hear a bird call that you do not recognize, it will be a Great Tit.

Contact calls

Contact calls are used to keep a pair, a family of fledglings or a flock in touch with each other. A chorus of piping and scolding notes heralds the approach of a flock of small birds. It is not easy to pick out and identify the callers as they flit from perch to perch among the foliage. The calls that accompany the flock are contact notes which serve to keep the birds in touch and maintain a heading in the same direction, despite the flock being scattered over a large area and frequently out of sight of each other. Individuals recognize the voices of their mates, parents and offspring but it seems that species in a mixed flock also recognize each others' species calls.

ABOVE: A chorus of call notes are often the first sign that a flock of Long-tailed Tits is coming through the garden. The calls help them to keep together as they forage through the foliage.

LEFT: As well as its song, the Garden Warbler has a number of calls, including ones for alarm and threatening rival males.

Signals

The songs and calls of birds are so familiar and part of our perception of them that their use of a sign language is easily overlooked. Displays, often used in conjunction with calls, play an important part in communication between birds. They are used mainly in aggressive situations and in courtship, although there is an overlap because courtship involves an element of threat when the pair are still uncertain of each other.

Aggressive displays are easiest to see at a feeder where birds are competing for food but they can also be seen around the garden in spring when male birds are establishing their territories. Fighting is time-consuming and dangerous, and displays get the message across without the birds coming to blows. They are postures which are the result of conflicting desires to attack and flee. A dominant Greenfinch or Chaffinch claims its place at a feeder or at scattered seeds underneath with the 'forward-threat' display in which it crouches, flutters its wings and threatens other birds with neck extended and open beak. The display indicates that the bird is strongly aggressive but the intent to attack is held back by the tendency to retreat. The reaction of the other birds is to retreat or show submission laced with some aggression, which they display by fluffing the plumage and retracting the neck.

ABOVE: *The Great Tit in the middle is raising its bill to show off its black breast-stripe as threat and a sign of dominance.*

Head up and wings open

Great Tits have two displays that can be seen at feeders. In the 'head-up' display the bird raises its bill to the sky to show off the black stripe running

ABOVE RIGHT: The size of the black bib on a male House Sparrow indicates its social status.

RIGHT: The white patches, surrounded by iridescent green, on a Woodpigeon's neck become more conspicuous when it cranes its neck. They act as a signal to other pigeons.

down its breast to the best advantage. It is used by males when defending their territories and it is mostly seen at feeders when the local territory-owner is claiming priority. The second, more eye-catching, display is the 'wings-open' in which the tit crouches with its wing partly spread and bill pointing towards its adversary. It is an aggressive display and is used to maintain the hierarchy in a flock. Bigger birds tend to be senior to smaller birds, older birds to younger birds, males to females, mated females to unmated females; with females taking a higher rank if they have a high-ranking mate. The width of the breast stripe indicates seniority: the broader the stripe, the higher the rank. By using the 'wings-open' display,

BELOW: A Great Tit tries to drive away a Starling in a 'supplanting attack' in which it attempts to land where the other bird is perched.

ABOVE: A Greenfinch prevents a Great Tit from landing on the birdtable by threatening with the 'forward posture'.

dominant tits make sure that they get precedence at feeders. Dominance does more than guarantee food. High ranking birds monopolize feeders near cover, while subordinates are relegated to those in more exposed places where they are more at risk from attack.

Territorial displays

Territorial displays are best seen early in the morning when the birds are not disturbed. They are common in early spring when territorial birds, especially Robins and Blackbirds, try to drive off intruders that refuse to flee. A territory-holding Robin perches near its rival and shows off its red breast to the best effect. If the rival is below, the owner stretches forward so the red breast is directed downward. If above, the owner stretches its bill skyward.

Courtship displays

Courtship displays are also seen mainly in spring, and they too are the result of one bird's desire to approach another bird, in this case for mating, but offset with a tendency to retreat from a fear of getting too close. Despite all the male's efforts to attract a female, his first reaction when she tries to join him may be to attack her.

Displays that result from this conflict can be seen performed by Blackbirds on the lawn or on the top of a wall. If it were not for the difference in the plumage of male and female it would be difficult to distinguish territorial from courtship displays because both involve aggression. Courtship displays may quickly change to aggressive displays.

As the pair gets to know each other the male becomes less aggressive. The signs of a settled pair include courtship feeding (p 93) and mutual preening. As well as cementing the bond, this has a practical advantage as it is often performed on the head and neck, which the bird cannot reach itself.

Bird society

'During the amorous season such a jealousy prevails amongst male birds that they can hardly bear to be together in the same hedge or field….It is to this spirit of jealousy that I chiefly attribute the equal dispersion of birds in the spring over the face of the country.'

GILBERT WHITE THE NATURAL HISTORY OF SELBORNE

The sociability of birds depends on the species, time of year and their situation. No bird likes close contact with its fellows, except in special circumstances such as mating and huddling in a roost. Birds like the Robin, Wren and Dunnock are solitary birds. Other species, like the Starling, Woodpigeon, Swallow and Long-tailed Tit are more social and spend much of their time in flocks. The Rook, House Martin and Heron nest in colonies. Social lives and the relationships between birds, whether defending territories or squabbling over food, are subjects that can be observed among the inhabitants of the garden.

ABOVE: *Throughout the winter, Fieldfares live in flocks for roosting, feeding and travelling.*

LEFT: *Birds of several species gathering at a birdtable behave mainly as individuals but interact when competing for food and co-operate in looking out for danger.*

Peck order

Many of the birds coming into the garden meet every day and will be familiar with each other. They form a small society. And, like any society, the members know their place in a hierarchy.

Every day there are Collared Doves on my lawn. Usually there is a single pair but they may be joined by two or three more to make a small flock. For most of the time they feed amicably but, once in a while, one bird lunges at another, which leaps out of the way with a flap of its wings. Both birds then settle down and start feeding again, but there is now more room between them. Similar altercations can be seen among the small flocks of Greenfinches and Chaffinches that feast on spilt seeds under the feeders. Squabbles are rarely serious and are often no more than a 'feint' in which one bird jabs at another, which promptly retreats, or one bird simply threatens, another with a display of aggression – the equivalent of raising a fist. See Chapter 6.

Equality is unknown in the natural world. The hierarchy among feeding birds is known as the 'peck order', a term that was coined during studies of the relationships within a flock of domestic chickens. A definite order of precedence at feeding time is established by fights when the chickens are first put together. At the head of the peck order is the dominant Chicken A which takes precedence over and can peck any other chicken. Chicken B may peck all others except Chicken A. Chicken C may peck all except A and B, and so on down to the poor subordinate chicken at the bottom which is pecked by all the rest.

The social order

Knowing their place in society helps birds because it brings stability and peace. Interruptions to feeding caused by arguments over precedence are kept to a minimum. Discipline is upheld by a quick peck at any bird that steps out of line. You can see social interactions between members of a flock when they descend on a feeder. Every now and then, one bird threatens another that moves away or the dominant

ABOVE: *Fights are mostly likely to break out between birds of equal social status because neither will retreat.*

bird may simply displace another in a 'supplanting attack'. I often watch Woodpigeons feeding on the clover that has replaced much of the grass on my lawn. They meander sedately across the sward, pecking as they go, and occasionally one wanders close to another. One of the two immediately recoils: it is the subordinate bird showing deference to its 'superior'. When there is a large flock in a field this behaviour is more obvious, especially when food, such as grain, is running short and competition builds up. Dominant birds are able to ensure that they get enough but the junior birds give way and go hungry. As supplies dwindle towards the end of winter, competition increases and the weak get less to eat. They become weaker and less capable of competing until they die. As with a nest of baby birds (see p103), when the going gets hard, the losers go to the wall.

Flocks

Many birds form flocks outside the nesting season after the family parties have dispersed. However, some birds are sociable all year and others are always solitary.

ABOVE: *Rooks are extremely sociable birds that live in flocks all the year round and visit the rookery even in winter.*

BELOW: *Despite the rhyme attached to numbers of Magpies, they are not often seen in more than pairs. Short-term gatherings occur during territorial disputes.*

Rooks, for example, nest in colonies and are sociable all year, while species such as the Robin and Great Spotted Woodpecker are always solitary (outside family life) and Wrens are sociable only when roosting. Redwings and Fieldfares spend the winter in flocks but their cousins, Song Thrushes and Mistle Thrushes, are not often seen in groups larger than family parties.

Flock values

I once walked along an overgrown hedgerow and Blackbirds flew out ahead of me every few yards. I did not start counting until too late but four-and-twenty Blackbirds is a good guess! I wondered whether these birds constituted a flock. This would assume that they were in contact with each other, by sign or sound, to co-ordinate their behaviour, despite being scattered along the hedge. If they were in a flock, it was a 'loose flock' unlike the dense masses of Starlings or Woodpigeons, but behaviour can be co-ordinated even if the birds are well spaced out. The tits that wander through woodland and occasionally into gardens are not so obviously part of a flock but the piping contact calls and the movement in a single direction show that they are as much a flock as a bunch of a dozen finches bouncing and twittering as they fly overhead. This leads to the question of why living in a flock is an advantage only to some species and only at certain times. The ornithologist Tim Birkhead has neatly summarised the value of flocking as 'finding food and avoiding becoming food'.

Many eyes

Birds that feed in flocks are usually looking for food that is patchily dispersed across the countryside, such as crops of seeds. A single bird could spend most of the day looking for somewhere to feed but its chances of finding a patch of food are much better if it can join other birds.

In winter, flocks of small birds forage among the bare branches. Most will be Blue Tits and Great Tits, perhaps with a Coal or Marsh Tit or a group of Long-tailed Tits, and there may be a lone Treecreeper, Goldcrest, Nuthatch or Chaffinch. The flock straggles slowly past, with the birds well spread out. They often lose sight of each other when they stop to search for insects and other tiny animals among the foliage or among the twigs and around the treetrunks. Then they hasten after their fellows, homing in on their calls. These flocks are most often seen in the morning when the birds are hungry after the night's fast or in bad weather when there is a special need to feed

well. As soon as they have finished feeding, the flock disperses.

More eyes, more food
Many pairs of eyes are better for spotting food and the finder will alert its companions. If one bird discovers that tearing off flakes of bark reveals insects sheltering underneath, it starts to tear off more flakes. The other birds note its success and follow its lead. Another advantage of working in a flock is that, when one source of food becomes exhausted, one or more of its members will have found another source of food and the rest of the flock takes note and quickly and efficiently changes

RIGHT: Small flocks of several species work their way through the trees. Many pairs of eyes help to spot sources of food and approaching danger.

ABOVE: *A flock of Black-headed Gulls finds an open space for resting and preening. One benefit is that there are many pairs of eyes watching for danger.*

BELOW: *Starlings spend most of the year in flocks but these split up for nesting and pairs defend small territories around their nesting places.*

tactics. These flocks are not so common in suburban gardens as in the countryside, probably because bird feeders supply easily located meals. The other main advantage in associating in flocks explains why birds of several species with different feeding habits join up to forage together. As will be explained in Chapter 13, another reason for flocking is that many pairs of eyes in a flock are better for spotting danger.

Eavesdropping

This system of eavesdropping on each other explains how flocks of birds, especially gulls and Starlings, appear apparently from nowhere as soon as food is put out. First one bird lands, then one or two more, and within a short space of time there is a milling throng. When I lived in the Scottish Highlands, the forestry workers told me how they often became the centre of attraction for a flock of Common Gulls when they stopped for a break in a countryside apparently empty of birds. There was much speculation about how the gulls gathered so quickly when the men opened their sandwich boxes. They were credited with supernatural powers of perception and communication but, as so often in cases of 'magic', the answer is sharp eyesight and quick reactions. If the men had looked around, they would almost always have seen at least one gull in view. Other gulls would be there, but so well-spaced over the hills and moors as to be out of sight. Yet the gulls were keeping a close eye on each other. When one suddenly dropped to the ground, its closest neighbours would suspect that it had found food and quickly fly towards it. Those farther afield would see the movement and follow suit. The result is that, within minutes, a crowd gathers. It might be a false alarm but the gulls behave as Russian shoppers were said to do: joining the back of a queue even if they did not know what there was to buy at the other end.

Territory

During the breeding season, rivalry between males, and sometimes females, increases. They stake out an area in which to breed. When Gilbert White of Selborne wrote about jealous male birds hardly bearing to be together in the same hedge or field, it was not an original thought.

Two thousand years ago Zenodotos of Ephesus had written 'A single bush cannot hold two Robins'. Perhaps Gilbert White was cribbing from the ancient Greek. The jealousy of Robins is expressed through songs and displays and Gilbert White knew that the object was to space birds 'over the face of the country'. The birds' 'space' is the territory which is simply defined as 'any defended area'. Bird territories range from a nesthole for House Sparrows and Starlings to a large feeding ground for tits and thrushes. The size of feeding territories depends very much on conditions, especially of food and the size of the population that needs to be accommodated.

Territory size

Bird territories are flexible in size. They provide space to breed and a sufficient area for feeding. If a habitat, such as a leafy suburb, is good and supplies plenty of food, more territories can be fitted in. Territory size also depends on the population. When numerous, the birds are crowded together. The areas given in the table are only an indication of the type of territory. Larger territories are used for feeding and small ones for nesting only. Territories may also vary through the year. Goldfinches nest in groups of a few pairs. During courtship their territories are about 250 square metres but contract to a small area around the nest of about 10 square metres and there can even be two or three nests in one tree.

The classic territory is one that supplies a bird and its family with all its needs: food, nesting place and roost. This is the kind of territory occupied by most garden birds, although gardens are not the best places to study territorial behaviour because they are too small and the territory of one bird may spread over several gardens. Territories are usually depicted as having boundaries as definite as fences, which gives the impression that the birds

ABOVE: *A Magpie defends its territory simply by perching in the top of a tree where it is clearly visible.*

SPECIES	AREA DEFENDED	
Tawny Owl	25 hectares	62 acres
Woodpigeon	4000 sq metres	4,800 sq yds
Great Spotted Woodpecker	5 hectares	12 acres
Swift	Nest only	
Starling	1 sq metre	1.1 sq yd
Wren	1 hectare	2.5 acres
Chaffinch	5,000 sq metres	6000 sq yds
Greenfinch	Nest only	
Goldfinch	250 sq metres	300 sq yds
Robin	1 hectare	2.5 acres
Blackbird	1 hectare	2.5 acres

BELOW: *A Great Tit spreads its wings and tail to threaten other birds that are coming too close for comfort.*

ABOVE: *Black-headed Gulls, in winter plumage, space themselves neatly along a rail, so they do not interfere with each others' space.*

keep within them as we and our neighbours keep within our gardens. And that they are just as hostile towards trespassers.

A flexible arrangement

In practice 'security' is often much more lax. Birds may leave their territories if they find a crop of fruit nearby and abandon them altogether in a spell of severe winter weather. In a garden where there is plenty of food this is most likely to result in an influx of birds. I have counted four cock Blackbirds and two hens feeding on my lawn, with another cock singing in a nearby tree. These birds were feeding harmoniously for the most part, and only occasionally was there a half-hearted chase. In fact, an amazing 72 Blackbirds have been recorded coming into one garden over the course of a single summer breeding season. They were arriving from distant territories in search of food and presumably they were attracted to this garden because it was a particularly bountiful feeding place. Individual Blackbirds seemed to prefer specific parts of the garden, which would reduce the chance of clashes. The resident cock must have felt swamped by numbers and decided that, since it could not beat the trespassers, it was better to join them. It decided that sharing the resources with other Blackbirds was preferable to spending all its time trying to chase them away while its own brood went hungry.

The outcasts

Not every bird manages to acquire a territory. It joins a population of outcasts or 'floaters' that keep a low pro-

file among the territory-holders or occupy a 'no man's land' in sub-standard habitat where there is a greater chance of death by starvation or predation. If floaters survive long enough, they may get the chance to take over a territory when its owner dies. It is a matter of being in the right place at the right time. When a territory becomes vacant, the take-over may be so quick that people do not realize that 'their' Blackbird or Robin has been replaced by another bird.

Setting-up a territory

A young bird can set up a territory only if it can find a space among the resident birds. The best place to insinuate itself is the loosely held ground at the junction of two or three territories. The alternative is to set up territory within an established territory. A young Robin trying to settle keeps a very low profile for several hours before starting to assert itself by singing and acting boldly. As it gains confidence, it spends more time singing until it has fully established its position. It may have to defend itself against the incumbent territory holder and if it behaves very boldly and is persistently aggressive it is likely to be attacked. The owner literally 'sees red' – the interloper's breast – and shows off its superiority by singing and displaying its own red breast. The newcomer retreats, before or after a scuffle, but there may be a real showdown. It is a myth that only humans

ABOVE: Two male Blackbirds dispute the boundary between their territories. Both are signalling aggression and a fight may break out.

commit murder. As many as 1 in 10 Robin deaths may be the result of border disputes in which the loser succumbs to pecks around the head. The violence stems from a territory being essential to a Robin's survival. It must risk a quick death to gain a territory or suffer a slow death by starvation if it fails.

Female robins

Robins defend territories throughout the year except when they are moulting after the nesting season. When the moult is complete in August they take up defence of the territory again, first throwing out any juveniles that may have set up home while they were absent. The difference now is that some of the Robins singing and chasing others from the birdtable are females. Females hold their own territories, even singing in their defence, and live a semi-detached life from their mates until they rejoin them to recreate the territories of the previous year. This may be soon after Christmas and the pair is soon amicably sharing crumbs again. Only snow and ice will drive them to seek dwindling food supplies elsewhere. The birdtable saves them from having to move.

Squabbling males and hen fights

Blackbirds are conspicuous birds and their territory disputes are easily observed. For several days, neighbouring males squabble over boundaries. They chase each other across the disputed area, taking it in turns to be the aggressor. Sometimes they come to blows, fluttering up breast to breast, rolling on the ground with wings flapping and feathers flying. Then they tear apart and retreat to safety, while giving raucous rattles of alarm. Hen fights are also quite common, and may be very violent. As territories are established, the threats and violence die down, and the borders are maintained by patrolling. If neighbours are residents from the previous year, there may be no need for any aggression. Patrolling is enough to restore the status quo.

Co-operating species

Unlike species that defend large territories with well-spaced nests, some species, such as Greenfinches, Linnets and Goldfinches are more sociable. A few pairs of birds will establish a cluster of small territories that provide no more than enough room for each pair to nest undisturbed by their neighbours. Feeding however takes place outside the territories. The advantage of this system is that the birds can co-operate in their search for food. Seeds are important food for their nestlings, but crops are often quickly exhausted, so it is very helpful to have several birds on the look-out for new sources. If food supplies fail then this can force the group to move to a new site to raise any succeeding broods.

BIRDS AND WINDOWS

Window tapping is a common occurrence. A bird, perhaps a Blue Tit, Chaffinch, Blackbird or House Sparrow, appears one day, tapping at a window pane. The habit can become obsessive and a Blue Tit once visited my study window daily, often returning at hourly intervals until, eventually, the habit wore off. It would tap persistently at the pane and even fly at it, banging its bill against the glass. The usual explanation is that the bird is a male that mistakes his reflection for a rival male invading his territory. This is not

always the case though as I have seen a female House Sparrow and Blackbird window tapping. The latter started window-tapping in mid-December and continued during some bitterly cold weather, which should have cooled any bird's ardour.

Stranger still are the reports of birds, usually crows, that smash against the glass so ferociously that the glass is smeared with blood and saliva. The attacks go on day after day, to the despair of the householders who have to clean up the mess and find that the birds are so obsessed that they squeeze through makeshift barriers to reach the window. There has never been a wholly satisfactory explanation to either window-tapping or the more extravagant window-bashing. I have not found any good reason why a bird should keep returning to seek out its 'rival' in a particular window.

Winter territory

The idea that birds are preoccupied with survival during the winter months is not altogether true where winters are mild. The garden goes quiet when nesting finishes and breeding birds go into retreat while they moult (See page 128).

For some species this is the end of territorial behaviour until the revival of the breeding urge next spring. They may move from the garden into the countryside or migrate abroad. Others like Starlings and Chaffinches change their behaviour and gather into flocks, but there are those that set up territories again before the winter. Choruses of song, aggressive interactions and snatches of courtship show that they have time and energy to think about the forthcoming breeding season.

As well as Robins, Blackbirds, Song Thrushes, Great and Blue Tits, Wrens, Dunnocks and Nuthatches are among the birds that set up territories in winter. There may be no more than a short phase of territorial behaviour in autumn. Then, as winter sets in, territories are abandoned when severe weather forces birds to travel farther afield in search of food. They return to their territories as early as January when the volume of song on a mild, bright day gives the impression of returning spring, and the boundaries of territories will be established by late February or early March, depending on the weather.

Not every territory is used for nesting. Mistle Thrushes form loose flocks after nesting but through the winter single birds, sometimes pairs, set up feeding territories. They defend crops of fruit such as holly, hawthorn, cotoneaster and ivy. Each thrush tries to conserve the crop on its tree while other food is available both by defending it from other members of the thrush family and itself feeding nearby. The tree will provide a useful source of food later in winter and early spring when other stocks are failing. This scheme breaks down, however, in severe weather when the Mistle Thrushes are overwhelmed by flocks of hungry Redwings and Fieldfares.

BELOW: Mistle Thrushes defend crops of berries so they will have a supply of food later in the winter when stocks are running low.

PART 3
Raising Families

Raising families, or passing on its genes to the next generation, is the object of every bird's life. Courtship, nesting and rearing the family, and for some species several families, takes up much of its time and energy during the spring and summer. The breeding tactics and strategies used by different birds have fascinated ornithologists but gardens are not a good habitat for nesting. Gardens are not usually good places for nesting unless nestboxes are provided. If there is a nest where it can be observed discreetly, watching the parents bringing off the young will be most rewarding.

ABOVE: Mallard ducklings are independent enough to feed themselves only a day after hatching. RIGHT: Juvenile Tawny Owls are still dependent on their parents for food three months after leaving the nest.

Pair formation

'For this was on seynt Valentynes day,
Whan every foul cometh ther to chese (choose) his make (mate).'

GEOFFREY CHAUCER: THE PARLEMENT OF FOULYS (FOWLS)

So wrote Geoffrey Chaucer in the 14th century, referring to the nice tradition that birds get married on February 14th. This is the time of year that bird song begins to fill the air and birds are seen flying about in pairs. It is a reasonable notion that birds should pair up at the end of winter and lay their eggs at the start of spring, but reality is not so simple. The resident birds that remain with us for the winter often start at least the preliminaries of breeding as early as the previous autumn.

The early bird gets the best mate

Robins, Starlings, Mallards and Rooks pair up long before Valentine's Day. They need to be ready to

LEFT: Blackcaps pair up as soon as the female (right) appears on the nesting grounds. BELOW: Mallards have a very visible series of courtship displays which bring male and female together.

start nesting as soon as the weather permits. Courtship starts at an early date and mates are chosen but there may be a long engagement through the winter and the union is not consummated until it is time to start laying eggs. It is not easy to follow the course of courtship. It may be so rapid that the birds settle together as soon as they meet or there may be protracted displays as either male or female spends time with several suitors before deciding to settle for one.

Fussy females

One function of courtship is to overcome the natural aversion of the birds coming into the close physical contact needed for mating. Females have to make the point that they are not rival males to be attacked and driven away, but recent studies of a number of common bird species have shown that there is more to obtaining a mate than two birds meeting and pairing. An unattached male may simply sing and display to court each and every passing female but the females are choosy about who to accept as a mate and a male may have to

ABOVE: A Woodpigeon keeps watch while its mate drinks. The pair keep separate from other pigeons during courtship and nesting.

court several females before one decides to accept his offer and stay. The reason for the male's lack of discrimination and the female's selectivity lies in the basic biology of the two sexes.

Natural selection

The object of a bird, or any kind of animal, is to leave as many offspring as possible to contribute to future generations, but male and female achieve this in different ways. Males can fertilize many females and potentially produce large numbers of young at little cost to themselves but females have to make a large investment in every baby. So a male's strategy is to go for quantity. He either mates with as many females as possible or ensures that his mate bears his offspring and no-one else's. Or, as we shall see, both. The situation is different for females. They go for quality. Their reproductive output is limited because they can lay only a small

ABOVE: Male and female House Martins work together to build the nest and the male may start the construction.

LEFT: A male Linnet guards his mate from the attentions of other males rather than help her gathering material for the nest.

number of eggs in their lifetime. Each egg represents a considerable investment in time and energy. The females' strategy is to ensure that the fathers of their eggs have good genetic characteristics to pass to the next generation and, in the case of most garden birds, that they will help rear the family.

Different agendas

It was the realization that each sex has its own, and perhaps conflicting, agenda that has enabled ornithologists to reveal the amazing richness of the private lives of birds – and some surprising goings-on in the garden. It is a charming sight to see two newly-paired birds flying about together. They seem inseparable. As soon as the female takes off the male follows hard on her heels so as not to let her out of his sight. But this is not an expression of simple devotion for his mate; rather the male is jealously preventing other males coming near his bride and giving her the opportunity to cuckold him. Mate-guarding, as this is called, is

obvious when a female finch is gathering materials for her nest.

A female Linnet started picking up stems of dead grass from my lawn and carrying them into the hedge in front of my study window. She was accompanied by her mate who often sang from a nearby perch while she was working on the nest. Earlier generations of naturalists would have interpreted this scene as the faithful, attentive husband serenading his industrious wife. The modern generation would notice that the cock Linnet did not carry a single blade of grass. His underlying emotion is jealousy rather than marital devotion. By shadowing his mate and singing, he is ensuring that other males are aware that he has exclusive access to her. As will be seen, he is well-advised to keep a close watch on his mate. (It should be noted, however, that males of some other species, such as Long-tailed Tits and Magpies, do help with nest-building but males of the finches, the true tits and the thrush family have opted out. See nest building, p 95).

Choosing a mate

It is a fairly new idea that female birds choose their mates rather than passively accept a suitor. Around a quarter of male Robins with territories fail to find a mate so there must be strong competition for females and they have the prerogative of choice.

Interestingly, research is revealing how birds select the most suitable mates. Choosing a superior mate allows a female to nest earlier and raise more or better offspring. The problem has been to understand how the female makes her choice. Apart from the size and quality of his territory (which includes food supply and nest sites), how can a male bird demonstrate his superiority? The female needs a character that can be related to his physical condition or experience. In Blackbirds, for instance, males with brighter orange bills are the most attractive. How its bill colour makes a male Blackbird a better mate is not known for certain, but these individuals are better fed and have fewer blood parasites, so they will be physically finer specimens. Female Great Tits seem to select males with broad breast stripes. It is known that these males make the best parents and bring more food to their nestlings but it is not known how plumage and parental behaviour are linked. Other birds use the male's song as an indicator of quality, as described in Chapter 6.

Monogamy

Monogamy is the rule for garden birds because neighbouring territories are roughly similar in quality in terms of food, nest sites and other resources. Only rarely is a male so 'well-off' as to attract an extra female. However, it would be better to talk of 'apparent monogamy' because bigamy by males has been recorded in many garden species, notably among Blue Tits. Usually both females lay in the same nest to make an outsize clutch. Male Starlings quite often pair with two females where nestboxes are close together but they help feed only the nestlings of their first female. Around one third of male Starlings are bigamous and one was recorded as mating with five females in one season! I was sent a strange, and apparently unique, example of bigamy in Blackbirds. It consisted of a double nest in which the two bowls not only touched but were physically joined by lengths of string and grass stems worked into both. I was told that it was built by two females which later sat side by side with a single cock in attendance. In the rare cases of

BELOW: Strange partnerships sometimes happen. Two female Blackbirds shared a mate and built this pair of joined-up nests.

MATE SELECTION

Older Chaffinch males obtain mates earlier in spring than young males because their singing and pursuit of females are more 'ardent'. It is reasonable to suppose that females recognize the potential of these males to make the most qualified mates. The best evidence for females' selection of mates comes from a study of Swallows. Female Swallows prefer males with longer outer tail feathers, the 'tail streamers' which are an aid to manoeuvrability. A Danish ornithologist, Anders Møller, has shown that older, and therefore presumably more experienced, males arrive earlier in spring and have longer tail streamers than younger, later arrivals. Proof of the effectiveness of long tail streamers was obtained by snipping 20 mm (¾ in) off the tail streamers of a selection of males and gluing the cut ends onto the tail streamers of another selection. Those with shortened tail streamers had difficulty finding mates, while those with lengthened tail

streamers mated more quickly, which gave their mates time to lay more eggs. Amazingly, females with 'long-tailed' mates reared twice as many young as females with 'short-tailed' mates.

Recent observations show that male Pied Flycatchers and Blue Tits with bright plumage are more attractive to females. The colour indicates that they are healthy and so are likely to be good providers. The same is probably true of 'ardent' Chaffinches and long-tailed Swallows. It is likely that, as common birds come under intense scrutiny, more examples of females choosing males because their physical condition is an indicator of their parental qualities will come to light.

ABOVE: *A male Chaffinch shows off to a female by turning sideways and displaying his white wing bars and red sides.* **LEFT:** *A female Chaffinch solicits the male's attentions by raising her tail and quivering her half-opened wings.*

bigamy in Blackbirds the two females are usually aggressive towards each other and divide the territory between them, but I also have a stranger record of a nest containing eight eggs with two hen Blackbirds taking turns to incubate with two cocks in attendance.

Polygamy

There are two species of garden birds in which polygamy is regular part of life. The male Wren is a most industrious builder of nests. It is quite a feat for such a small bird to build a domed structure of leaves, grass and moss, measuring about 15 cm (6 in) across, but the Wren builds three or four nests per season, and some males may construct ten or more. The female's contribution is confined to adding a lining to the nest she selects for her eggs. The surplus

nests are called 'cock's nests' and they are left to collapse through the action of the weather or questing nest robbers. It may seem that Nature is being surprisingly uneconomic with male Wrens wasting an enormous amount of time and energy in unproductive labour, but there is method in her madness. Male Wrens that build many nests are more likely to attract females to settle with them. The nests are the equivalent of the Swallow's long tail-streamers and the Starling's elaborate song, except that the number of nests a Wren builds is related to the richness of the habitat in its territory and hence food supply, rather than the physical condition of the bird.

ABOVE: Jackdaws mate for life and rarely a male mates with two females. Both lay eggs and all three help rear the nestlings.

LEFT: House Sparrows pair for life but this does not stop females having casual relationships with other males.

Extra pair paternity

A male bird does not have to form a partnership with a female to father at least some of her offspring, because it turns out that male birds' efforts to guard their mates from the attentions of interlopers are surprisingly ineffective.

Genetic fingerprinting gives proof of parentage and a study of Blue Tits showed that nearly one half of the nests contained nestlings which had been sired by extra-pair matings: the male feeding them was not their biological father. One brood of Starlings was fathered by three different males. The explanation for the females' infidelity may be that she is seeking a male that is superior to her mate to father at least some of her offspring, while not losing the mate that will help her raise the family. Cuckolding often takes place when the female leaves the nest in the morning, the time when she is at her most fertile and she will seek

out better-quality males. It is not surprising to find that Swallows with long tail streamers are more successful at fathering extra young and that the offspring of these fleeting unions with better males have improved chances of survival.

The devious Dunnock

No bird has shown more how wrong we can be with interpreting behaviour than the Dunnock. Nick Davies, who studied Dunnocks in the Cambridge University Botanic Garden, quotes the Reverend F. O. Morris who wrote that the Dunnock 'exhibits (in deportment and dress) a pattern which many of a higher grade might imitate with advantage to themselves and benefit to others through an improved example'. The Dunnock has a sober dress, indeed, of grey and brown, and it can be seen in any garden industriously searching for tiny insects, often in the company of its mate with which it will co-operate to rear its family in a manner that would gladden the heart of any Victorian moraliser. Or so it would seem. Davies' observations revealed that the female Dunnock has made an art of extra-pair mating that makes its private life rich even for a TV soap opera. Her strategy is based on the fact that a male Dunnock will help feed the young of any female he mates with.

BELOW: *The Dunnock has an interesting social life. A female frequently mates with males other than her mate. The benefit is that they will help rear her family.*

At the start of the breeding season male Dunnocks set up territories and females set up independent feeding ranges. If a female's range overlaps a male's territory, she pairs up with him monogamously. But if her range overlaps two territories, she mates with both males. This arrangement provides the female with a large range – useful when food is scarce – and the overlap gives her two males to help feed her nestlings. The two males learn to accept each other's presence, although one remains dominant and tries to guard the female from the other. She, however, will try to evade the dominant male and mate with the subordinate, or even fly through the undergrowth in search of strangers. Once they have mated with her, they are bound to help feed her nestlings because they could possibly be their own. From her point of view, an extra 'uncle' or two will ensure that her babies are well-fed and will grow up strong.

SPARROW PARTIES

An outbreak of frenzied chirping draws attention to a group of House Sparrows having a free-for-all on the ground. They take-off and disappear into a hedge, still calling vociferously. The members of this lively group are so taken up with each other that they take no notice of human bystanders and may even fly into them. These gatherings have been called 'sparrow parties' or 'sparrow weddings'. It is not easy to check but they consist of several males but only a single female.

These parties are seen in spring, before egg-laying. They start with one male courting the female which attracts other males to the scene. The female tries to fly away but is followed and she has to fight to defend herself against the males, which are trying to court her with wings shivering, tails raised, breast fluffed out and much chirruping. Her mate tries to defend her and gives alarm calls but does not always succeed. The result is that the female's clutch will contain eggs not fertilized by her regular mate.

Courtship feeding

Courtship feeding is a ritual that can be seen in the early part of the breeding season before nesting has started. It is more frequent after the initial courtship and pairing, when the pair has become firmly attached and has settled into nesting.

The male brings gifts of food to the female, thereby helping to redeem his poor reputation in those species, like the finches and tits, where he does not help with nest-building. At first sight it looks as if an adult is feeding a fledgling because the female accepts the offerings or even solicits them by begging like a juvenile, 'with a fondling tremulous voice and fluttering wings' as Gilbert White wrote of the female Rook.

Necessary nutrition

The food provided by the male is a significant addition to the female's diet during egg-laying and incubation. This is the time when she is in need of extra nutrition to form the eggs or to sustain her while she is tied to the nest. A female Blue Tit gets enough food from her mate to manufacture the eggs so she needs to forage only to maintain herself at this crucial time when the weight of the developing eggs in her body makes flying difficult. During incubation, courtship feeding allows the female to stay on the nest as long as possible to prevent the eggs cooling and keep nest robbers at bay. So the male's behaviour directly helps the survival of his offspring. However, courtship feeding sometimes has a purely symbolic side. Female gulls have been seen coming back from feeding and soliciting food from their hungry partners! It seems that the action has more to do with cementing the pair bond. The pigeon family has taken this further in a ritual called 'billing' in which no food passes. The pair merely touch bills or the female puts her bill into the male's open bill.

BELOW: *A male Rook feeds his mate. She solicits the meal in the same way as a young bird begging for food from its parents.*

ABOVE: Woodpigeons have a symbolic form of courtship feeding called 'billing'. No food is passed but the bond between the pair is reinforced.

RIGHT: Courtship feeding by Robins is a common sight in gardens and is a preliminary to nesting. The adults will later be seen feeding the fledglings in the same fashion.

FIDELITY AND DIVORCE

It is easy to assume that the pair of tits visiting the nestbox, the Song Thrushes hopping together across the lawn or the two Bullfinches stripping the apple blossom are the same two that were visiting the garden last year, but this is probably not the case.

Mortality of small birds is high, with only about one third of adult Blackbirds and half the tits surviving the winter (see p131). Most of the winter's survivors will have to find a new mate in spring.

If both birds do survive, they may pair again; not because the bond between them was retained, but because they remain in, or return to, last year's nesting place and naturally come together again. About three-quarters of Robins return to the previous year's mate but they are being faithful to the site not the bird.

Even if both of the pair have survived, divorce is a possibility, especially if the previous year's nesting attempts were unsuccessful. It is better to try your luck with a new partner than risk another failure with the old one. Carrion Crows and Rooks may divorce when a pair has failed to breed successfully but Jackdaws are parted only by death. With Blue Tits, the divorce rate can be as high as 85 per cent, probably as a result of the females looking for new partners, but female Magpies divorce to find a territory with better resources rather than a better mate. Desertion by male birds is less likely because they dare not lose their territories.

Nests and eggs

'Birds' ways with their nests and eggs illustrate Nature's infinite variety. The female lays the eggs. This is, really, the only general statement that can be made on this subject without reserve or qualification'.

VISCOUNT GREY THE CHARM OF BIRDS

Viscount Grey was writing about the whole range of British birds but most of the birds likely to nest in gardens conform to what we think of as normal nesting behaviour. The nest is a basket of twigs, grasses and other plant materials, built in a tree or a nestbox. The eggs are laid and incubated by the female with the male in attendance and both parents feed the young in the nest and care for them when they first take to the air. The Garden Nesting Survey organized by the British Trust for Ornithology has recorded 60 species nesting in gardens but only a few common species – notably House Sparrow, Starling and Blackbird – are regular garden nesters, other than in nestboxes. Gardens are not particularly good places for nesting. They are usually not well-endowed with good nest sites, unless nestboxes are provided, and they do not provide the food base required to raise families, unless well planted with trees and shrubs. Yet a single pair of nesting birds provides plenty of interest in both routine activities and some unexpected turns of behaviour.

LEFT: Goldcrests often nest in gardens where there are conifers in which to sling their delicate nests of mosses and lichens, bound with cobwebs and lined with feathers and hair.

Nest building

The first sign that nesting is about to get under way is the sight of birds prospecting for nest sites. Tits start to examine nestboxes during the previous autumn but birds nesting in open situations can be seen moving through the undergrowth in spring.

They are not feeding but examining the branches and twigs in a purposeful sort of way. How birds choose where to start building a nest is something of a mystery and even a nestbox will be ignored for years and then occupied overnight when it is moved to a new site. Sometimes the bird gets a little confused. David Lack described how a pair of Robins built in a stack of drainpipes: 'They placed nests or parts of nests in twenty-three of the pipes.' Eventually they settled in one pipe, completed the nest and laid their eggs.

The Choosy Chaffinch

A female Chaffinch prospecting for a nest site flies among trees and bushes, landing on branches and hopping down to inspect crutches where two or three branches sprout. She turns round and about, flicking her tail and inspecting the site as if sizing up its potential. She may bring some building material before deciding that the place is not suitable and continuing her search. Eventually she makes her final decision and starts to collect nest material industriously. She must have recognized some con-

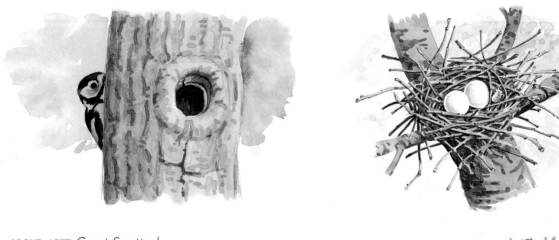

ABOVE, LEFT: Great Spotted Woodpeckers use the same hole in a treetrunk in successive years.
ABOVE, RIGHT: The Collared Dove builds a frail platform of twigs for its two eggs.
BELOW, LEFT: A Wren's nest is a hollow ball of leaves hidden among foliage.
BELOW, RIGHT: Magpies build a conspicuous dome of twigs that protects the eggs and nestlings from predators.

figuration of branches that will give a good support for the nest. If only we knew what this was, we could more easily lure Chaffinches and other birds into nesting in the garden!

Secret nesting

Unless the nest is in a safe place, like a tit's in a nest-box or a House Martin's nest that fits snugly under the eaves, nest building is a hidden activity, The sign that building is in progress is the industrious gathering of materials: House Sparrows flying under the eaves with trailing grasses, Blackbirds gathering beakfuls of mud from the edges of ponds and puddles, and Goldfinches gathering cobwebs from walls and fences. Disturbance at this stage must be avoided. Wait for the nesting season to finish and the leaves to fall. The old nest can then be inspected at leisure. A search in winter through bare hedges and bushes reveals the remains of nests, probably of Blackbird, Song Thrush and Chaffinch but perhaps Goldcrest or Long-tailed Tit.

ABOVE: A Rook tugs at a twig to break it from the tree. It has lost its balance as the twig snapped.

LEFT: Blackbirds sometimes nest in very exposed situations, like window sills. Despite this failing, they are successful breeders in gardens.

Incredible Creations

It is amazing that birds build such intricate cradles with only the bill as a tool. I used to have a Bullfinch's old nest which I had removed from a privet hedge. It was a tiny, delicate bowl of rootlets. It seemed so fragile that it was a wonder it had not burst apart as the family of nestlings grew, but the materials had been woven into a flexible yet strong fabric.

The most intricate nest of all is the bag-shaped nest of the Long-tailed Tit. Made mostly out of mosses bound with spiders' web, it starts as a bowl which is roofed over, then lined with hundreds or thousands of feathers and decorated on the outside with countless flakes of lichen. Building takes a fortnight and the two tits travel an incredible 600–700 miles (960–1130 km) gathering material. The construction is even more remarkable for housing a full-grown brood of eight to twelve nestlings.

A different perspective

One reason why a bird's nest seems so marvellous to human eyes is its diminutive size. Richard Jefferies, the Victorian essayist, put his finger on it over a century ago: 'To a man or woman, so many times larger than the nest, its construction appears intricate… It does not look tiny to the bird. The horsehair or fibre, which to us is an inch or two long, to the bird is a bamboo or cane three or four feet in length. No one would consider it difficult to weave cane or willow wands as tall as himself.' Jefferies continued, 'To understand birds you must try and see things as they see them, not as you see them.' This is good advice for anyone intending to study animal behaviour.

The building process

Starting the nest must be the most difficult part of the building process. Chaffinches begin by wrapping threads of spiders' web around branches to make anchor points. Grass and moss are added, with more web to bind them, until a firm pad makes a foundation for the nest cup. One can ponder on the importance of spiders in the lives of small birds, not simply as food but as providers of vital nesting

ABOVE: *Unlike the nests of other tits which are hidden in holes, the feather-lined nest of the Long-tailed Tit is very vulnerable to predation.*

material! A rookery is as good a place as any for watching the nest-building process. Sticks are laid in a fairly random way in the chosen site among the branches. Whether they remain in place or fall to the ground seems to be a matter of chance. When enough have accumulated, the Rooks start to push new sticks into the growing pile to make a solid foundation.

The final straws

The interior of a nest, taking a Song Thrush's or Blackbird's mud-lined cup as an example, is so round and smooth it could have been made on a potter's wheel. I had the opportunity to see how this is done when a Blackbird started to build on a window ledge. (Blackbirds often nest in ridiculously exposed places.) She amassed a small pile of dead

MUD NESTS

Starting a nest is difficult even for Swallows and House Martins, which make nests of mud. They gather mouthfuls of mud from the edges of ponds or puddles near the nest site and plaster them in place. The brickwork under the eaves where these birds nest are peppered with small lumps of plaster where initial building attempts have failed. Eventually a foundation is formed by pressing successive mouthfuls of mud into place. The birds take care not to work too quickly when the mud is wet or the structure will disintegrate.

BELOW: *Nuthatches reduce the size of the opening of their nestholes by plastering them with mud.*

ABOVE: *A muddy edge to a pond or even a puddle may attract nest-building Swallows and House Martins.*

Counting visits per hour shows that activity is greatest at the start of the day and slows down in wet weather when the birds have to spend more time searching for insects for food. Rootlets, hair, grass and feathers are worked into the structure to reinforce it and, when a nest eventually falls down, its composition can be inspected.

grass stems and then hopped round and round among them until she had a neat ring of woven, or more correctly felted, grass. There are two basic movements in nestbuilding: scrabbling, in which the bird presses down and pushes back hard with each foot alternately; and turning, in which the bird revolves in the sitting position. The sides are built up by lifting loose material onto the rim and tucking it in with the bill. These simple actions are all that is needed to work the mass of nesting material into a symmetrical cup that will comfortably hold the eggs and sitting bird.

Hidden handiwork

Concealment is important and early nests built before the leaves have sprouted are most likely to be lost to predators. An ornithologist told me he thought that the Blackbirds in his garden had failed to nest after Magpies settled in his neighbourhood. A patient search revealed that there were still plenty of nests, but they had been built in dense undergrowth instead of exposed sites. Nest-builders try to approach the site unobtrusively (not easy when trailing a length of grass). I watched a Blackbird collecting moss and mud from my pond to line her nest somewhere in my neighbour's shrubbery. On each journey she paused on our border fence to see if she was being watched. Then she flew into a hawthorn tree and perched for a minute or two before slipping out of sight into the foliage. The stealthy, roundabout approach to her well-concealed nest should deceive the pair of Magpies that were keeping watch from a vantage point in a tall tree. She would soon be putting all her eggs in one basket and she had to hide its location from the robbers' prying eyes.

Not for re-use

Despite spending so much time and energy building their nests, it is unusual for garden birds to use them for a second clutch. Rooks and House Martins refurbish their old nests, but the general rule is that nests are abandoned after use.

There are exceptions, such as a Blackbird that reared its families in the same nest for three years in succession. Blackbird nests, with their mud linings, are more substantial than most and a little renovation makes them as good as new. Nests in sheltered positions are more likely to be used again, perhaps because they are less weathered, but most birds may work to a rule of thumb that it is safer to start a new nest than to risk the collapse of an old one.

Another good reason for building a fresh nest is to keep parasites at bay. A nest may harbour hundreds of blood-sucking lice, mites and fleas, which lie dormant during the winter and infest the next year's occupants. Great Tits using infested nestboxes are more likely to desert and their eggs hatch less successfully. Incidentally, if you think it is a shame to empty a nestbox of nest material which took a bird days of work to gather, and which could perhaps make a nest for bumblebees next year, there is a solution. Remove the nest from the box, microwave it to kill the fleas and put it back in the box. Clean the box thoroughly as well because parasites lurk in crevices.

A fragrant home

It has been known for a long time that Starlings line their nests with flowers and fresh leaves. They are not the only birds to do this and the habit is puzzling. However, it has been discovered that some plants collected by Starlings have insecticidal properties. Tests have shown that the population of blood-sucking parasites, such as the fowl mite, can increase a hundredfold if green leaves are removed from Starling nests. Plants selected by Starlings include agrimony, yarrow, parsley, mint and, not surprisingly, fleabane – traditionally used for killing insect pests. Blue Tits have been seen fetching leaves of lavender, yarrow and fleabane. What is more, as the leaves in the nest lost their aroma, the tits replaced them with fresh material.

ABOVE: *The Little Owl may use the same nest hole for several years but it is more likely to change from year to year.*

BELOW: *Starlings solve the problem of harmful parasites by lining their nests with aromatic plants.*

Incubation

After the nest is complete there may be a pause before egg-laying starts. Small birds lay an egg a day, usually in the morning, and incubation does not start until the clutch is complete.

Birds prepare for incubation by shedding belly feathers. The bare patch of skin develops enlarged blood vessels and acts as a 'hotpad' for keeping the eggs warm. The presence of a brood-patch is a sure sign that a bird has been incubating. This is something that can be checked when examining a dead or injured bird, except in the case of pigeons and doves, which have brood-patches all year, and

CLUTCH SIZES, BROODS, LAYING SEASONS AND INCUBATION PERIODS OF SOME GARDEN BIRDS

Species	No. of eggs in clutch average/ min–maximum	No. of clutches laid	Approx. main laying period	Incubation period (days)
Collared Dove	2	3–6	early March	14–18
Tawny Owl	2–5 (1–6)	1	end February	28–30
Swift	2–3 (1–4)	1	late May	19–20
Great Spotted Woodpecker	4–7 (3–8)	1	late April	10–13
Swallow	4–5 (2–7)	2–3	early May	15
House Martin	3–5 (1–7)	2	early May	14–15
Wren	5–8 (3–9)	2	early April	16
Dunnock	4–6 (3–7)	2	late March	12–13
Robin	4–6 (2–8)	2	late March	14
Blackbird	3–5 (2–6)	2–3	early March	13–14
Song Thrush	3–5 (2–6)	2–3	late March	13
Goldcrest	9–11 (6–13)	2	early May	16
Spotted Flycatcher	4–6 (2–7)	1–2	late May	12–14
Long-tailed Tit	8–12 (6–15)	1	late March	13–15
Blue Tit	8–13 (2–18)	1	late April	14
Great Tit	3– 18	1–2	late April	14
Nuthatch	6–8 (5–12)	1	early April	15
Magpie	5–7 (3–10)	1	mid April	21–22
Starling	3–5 (2–10)	1	April	12
House Sparrow	3–5 (2–7)	2	April	11–14
Chaffinch	4–5 (3–6)	1	May	12–13
Greenfinch	4–6 (2–7)	2	May	13

There may be some variation in the number of broods, laying dates and incubation periods. Many birds lay a replacement clutch if the first is lost.

Source: Handbook of the Birds of Europe, the Middle East and North Africa: Birds of the Western Paleactic; Cramp, S., Snow, D., Perrins, C. Oxford University Press.

ducks, which never develop proper brood-patches but pluck some of their belly feathers.

Simultaneous hatching

A bird may sit on its nest before the clutch is complete but most common birds do not start to incubate until all the eggs have been laid. (Cold eggs are therefore not always a sign that a nest has been deserted.) Incubation of the entire clutch starts at the same time so the eggs hatch more or less simultaneously and the fledglings leave the nest more or less together. Of common birds, only owls, birds of prey, herons and gulls start incubation with the first egg. As a result, the eggs of these birds hatch in sequence and there is a distinct range in size from the oldest to youngest nestlings. If there is a food shortage the youngest die first, giving the older ones a better chance of survival.

The great unknown

The nesting period is frustrating for the watcher. It is the most interesting, action-packed part of the bird calendar but most of what is happening has to remain unseen because of the dangers of disturbing the birds or revealing the location of the well-hidden nest to predators. Something of incubation behaviour can be seen when birds are using nestboxes or have nests in exposed sites, like House Martins, Spotted Flycatchers and Swallows. Occasionally Blackbirds, Collared Doves or Mistle Thrushes nest in exposed places and the comings and goings of the parents can be watched.

A mostly female occupation

Most garden birds leave the incubation of the eggs to the female, but it is shared by both sexes in Starlings, woodpeckers, House Martins, Swifts, and the pigeon family. There is more to incubation than simply sitting on the eggs. The bird constantly monitors the temperature of the eggs and adjusts its behaviour, by sitting more closely in cold weather, for instance. The incubating bird does not spend its entire time on the nest. It leaves at intervals to feed, defaecate and preen. In bad weather it has to spend more time feeding so there is a danger that the eggs will chill and die.

CUCKOO TRAITS

One morning I found three Starling eggs lying on the lawn under a nestbox. They were intact, so robbery by a predator could be ruled out. They had been ejected from the nestbox by the female Starling, but she had not been as bad a mother as this suggests. Quite the reverse. Some Starlings lay eggs in the nests of other Starlings and save themselves the trouble of rearing their offspring by getting another bird to do it. This is the same strategy that the Cuckoo has brought to perfection. However, the Starling's ruse becomes unstuck if the eggs are laid before the rightful occupant of the nest starts to lay her own clutch. She recognizes that the foreign eggs do not belong to her and throws them out. To be successful, 'egg dumping' as it is called, has to be carried out after the rightful occupant has started her own clutch. She cannot then tell the strange eggs from her own and she dare not risk throwing her own eggs out by mistake. So the 'cuckoo' Starlings are raised as part of her family.

A dilemma

Incubation is a demanding task because the sitting bird has to use its own body heat to maintain the eggs at a high, even temperature, yet its time for feeding is severely restricted. This is a problem especially for tits, finches, thrushes and other species in which the female alone incubates, although they are assisted with food brought by their mates. You can see female tits being fed when their mates approach the nestbox. They emerge and flutter their wings to encourage the males to pass over food. Bad weather may cause the loss of clutches through cooling because the sitting birds have to spend more time off the nest in the search for food. A prolonged spell of cold or wet weather may make feeding so difficult that birds are forced to abandon the nest and let the eggs or nestlings die.

Rearing the young

'With the hatching of the eggs and the advent of the young the cares and labours of the parents enter a new and more exacting phase. Where the female has been left to carry out the work of incubation alone she is now joined by her mate, no longer able to ignore his responsibilities, or perhaps roused into activity by the sight of his offspring.'

W. P. PYCRAFT A HISTORY OF BIRDS

As with human families, life for parent birds changes with the arrival of the young. The parents continue to brood the nestlings for several days after hatching but spend increasingly longer spells off the nest. In those species in which the male plays no part in incubation (most garden birds) he now plays a very active part in caring for the nestlings. The newly-hatched nestlings are unappealing – naked, blind and little more than a bag for receiving food. Two simple instincts govern their lives: begging for food to be inserted at one end and voiding the waste at the other. In the course of two or three weeks they develop into fully-grown birds that can fly, but they usually need some care from their parents before they are fully independent.

SOME BASIC TERMS

A young bird is often called a *chick* before it can fly but this word should be used only for young birds that leave the nest and run around soon after hatching, such as Chickens (the prototype chicks), Mallards, Moorhens and Pheasants. Young birds that stay in the nest until they can fly are called *nestlings*, although ornithologists often refer to them as *pulli* (singular – *pullus*). The eggs in the nest are *incubated* and the newly-hatched young are kept warm by *brooding*. When a young bird has grown its first set of feathers and flies for the first time (and leaves the nest if it is a nestling), it is said to have *fledged*. It is referred to as a *fledgling* for a few days, then it becomes a *juvenile* while it is still in its first plumage, which may be different from the adult plumage. The time from hatching to fledging is the *fledging period*.

LEFT: *The downy ducklings of Mallards are chicks rather than nestlings because they leave the nest soon after hatching.*

Feeding nestlings

It is over 50 years since my father showed me a nest of newly-hatched Yellowhammers in a hedge. It was a good lesson in the simplicity and economy of instinctive behaviour.

The tiny nestlings were asleep in the bottom of the nest but when my father tapped a twig which formed a support for the nest, they sprang into life. The thin, scraggy neck of each nestling was thrust upwards and the bill opened wide to reveal a flesh-coloured mouth with a yellow border. With all the nestlings gaping at once, it looked as if a bunch of flowers had suddenly blossomed. The nestlings were reacting to my father's tap as if to the arrival of a parent carrying food. The brightly-coloured mouth, which is a feature of all nestlings, is a signal that they are hungry and an indicator of where the parent should put the food. A few days later, the yellowhammer nestlings' eyes had opened and they now directed their gaping towards a moving object, as demonstrated by my father's moving finger, but more naturally the arriving parent.

ABOVE: When they are first hatched and their eyes are closed, Blackcap nestlings gape for food automatically when they feel the vibration of a parent arriving.

Hungry mouths

The parent stuffs food indiscriminately into the nearest of the gaping mouths, but there is a simple mechanism that ensures all the nestlings are fed in turn. As one nestling becomes replete, it gapes less strongly and eventually fails to respond to the parents' arrival. Meanwhile, the hungrier nestlings continue to beg and receive food. Thus all the nestlings are fed in turn and, if food is plentiful, the parents get a respite while the brood sleep off their meal. In practice, parents of large families have difficulty finding enough food to satisfy all their nestlings. When bad weather makes foraging difficult, the strongest get the lion's share of food and thrive while the weaker ones succumb. By our standards this is unfair, but in the long run, it is better for the parents to rear a few strong youngsters than many half-starved ones.

BELOW: When they are older and they can see, the nestlings direct their gaping mouths at the parent bearing food.

Waste removal

After delivering the beakful, the adult bird waits a moment for one of the nestlings to turn, present its backside and void its droppings bound in a gelatinous package. The parent picks it up in its bill and carries it away. It is easy to see birds carrying these white faecal pellets away from the nest and dropping them at a distance. It helps keep the nest clean and also prevents the position of the nest being given away by telltale splatterings of 'whitewash'. House Martins, whose nests have no need for camouflage, do not remove droppings after the first four days and the walls and windows beneath the nests become liberally soiled – to the annoyance of tidy householders.

ABOVE: Life is considerably easier for a parent Robin when its young leave the nest and follow it around the garden.

BELOW: A Long-tailed Tit drops food into the nearest, hungriest nestling. As each one becomes replete it is replaced at the nest entrance.

Nestling nutrition

As I related in Chapter 1, the invalid naturalist H. J. Massingham whiled away the time by watching a pair of Blue Tits feed their family. It is easy to watch parent birds flying to their nests, in some species with insects or worms dangling from their bills. Sometimes they pause on a perch before diving into the nest and it is possible to identify the kind of food being brought in. Blackbirds and Song Thrushes have a preference for earthworms, Starlings for leatherjackets, and Blue Tits and Great Tits for caterpillars, while even the seed-eating finches supplement their nestlings' diet with insects.

It takes a round figure of 10,000 caterpillars to raise a brood of

Great Tits to fledging but, with smaller animals also on the menu, there are likely to be many more visits to the nest in the course of the two or so weeks that most garden birds take to rear a family. When food is in short supply, the parents struggle to satisfy the demands of the young. Sometimes they fail and the weaker ones are sacrificed, as explained above, unless alternative supplies are found.

Emergency measures

I once watched a pair of Blue Tits during a thunderstorm. They were visiting a nestbox, where shrill cries indicated a hungry brood. I was interested to see how they would fare, because Blue Tits rely on leaf-eating caterpillars as food for their young and heavy rain washes caterpillars off the trees. My Blue

RIGHT: A Blue Tit keeps the nest clean by removing droppings from one of its nestlings. It will drop them at a distance from the nest.

EGGSHELL REMOVAL

Sitting birds are fastidious about their nests. As well as removing droppings before they pollute the nest, they take away any stray litter and generally keep the inside of the nest tidy. This includes empty shells, which are carried away and dropped a distance from the nest, where they will not give away the exact location of the nest to predators. The reason for removing empty shells is believed to be to prevent them slipping over other eggs or nestlings and so harming them.

It is not unusual to find an empty shell on the lawn. The neat shape, like a boiled egg that has had its top removed, shows that it has not been eaten by a nest robber. The shell is proof that a bird is nesting nearby and its appearance gives the date of hatching. This is one of the few times when birds' eggs can be handled to see their exquisite form and colouring and examined for identification.

peanut from the nearby feeder. It had never done this before and was a sign that natural food was hard to find. Stocking birdtables and feeders through summer used to be frowned upon because they were 'unnatural' foods and whole peanuts could choke nestlings. Summer feeding is now encouraged because it has been realised that birds nesting in gardens can easily run short of natural food even in good weather.

The best food available

The ability to choose suitable food should help to allay worries that putting out food for birds in spring and summer may lead to them giving inappropriate food to their youngsters. Apart from concern that nestling tits could choke on peanuts (which incidentally is a very rare occurrence and can be prevented by using feeders with rigid wire mesh) they are sometimes thought to be an inappropriate, 'artificial' food. However, I, among others, have watched parent

ABOVE: Watching tits visiting their nestboxes gives the best opportunity for observing the nesting habits of birds. It may be possible to see what food they are bringing to their nestlings.

Tits had changed to bringing beaks full of flies and other small insects for their young. These take longer to collect and are less nutritious than caterpillars, so are a form of emergency food.

It was also significant that one Blue Tit sometimes spent a few minutes bringing fragments of

tits taking insects to a nestbox but breaking off to have a quick, 'fast-food' snack at the feeder, before resuming foraging for insects for the nestlings. It may be that peanuts are only fed to the nestlings when 'natural' food is scarce. I have watched a male Blackbird rearing the family on his own after a car had killed his mate. He once took a crust of white bread to the nest. This was wholly inappropriate but may have been a despairing effort by an overworked parent struggling to feed his family in dry weather when worms were scarce.

Leaving the nest

As spring turns into summer, the garden becomes busy with new families of parent birds followed by their youngsters. The squeaky calls of fledglings waiting patiently or actively chasing a parent and begging for a meal is the first sign that young birds have left their nests.

We say that our children have 'flown the nest' when they have left the parental home for an independent life but this is not strictly accurate because the fledglings of most birds continue to be supported by their parents during the critical time after leaving the nest.

The first flight

When the young are ready to leave the nest, they may be encouraged by the parents calling from a nearby perch or bringing food to the nest then backing away to draw them out. Fledglings are easily recognized by their absurdly short tails. Their wing feathers are also not fully developed and their flight looks clumsy and laboured. They spend their first few days sheltering in the undergrowth and flying only short distances, while their parents bring food to them. Although able to fly instinctively and keep airborne as soon as they launch themselves, landing is more difficult and requires practice for perfection; which is where many young birds come to grief. I watched a House Sparrow attempt to land on a wall, miss its footing and flutter out of control in a nose-dive for 6 metres (20 ft) before recovering and flying back to the perch.

Dangerous days

The first days after leaving the nest are the most dangerous time in the life of any bird. Many newly-fledged birds succumb at this stage. Their clumsy flight and lack of manoeuvrability makes them easy prey for predators. It is also vital that the newly-fledged youngsters maintain contact with a parent so that they will continue to be fed

ABOVE: Despite growing-up in the cramped confines of the nestbox, a young Blue Tit knows how to fly as soon as it launches itself into the air.

BELOW: When a juvenile Blackbird is learning to fend for itself, it is helped by supplementary meals from its parents.

HOUSE MARTINS TAKE TO THE AIR

The antics of House Martins when their young are leaving the nest can give the impression that they are teaching them to fly. This is quite wrong; the ability to fly is instinctive. The first flight is an incredible step when we remember that this is the first time the young martin has been able even to spread its wings. It is as if we could ride a bicycle at the first attempt. Birds that grow up in enclosed nests do not even have space to spread their wings fully and have to produce the complicated wing movements from scratch.

House Martins are unusual in the amount of attention given to the nestlings when they fledge. When the time for the first flight arrives, the nest may attract a flock of martins, which fly to and fro chirping excitedly. Every now and then one flies up to the nest and appears to touch bills momentarily with one of the nestlings peering out. Eventually, one martin, usually the mother of the family, lands at the nest and then backs off slowly, while the youngster leans further out, and then launches itself boldly after her.

Sometimes House Martins seem to get the timing wrong and the young come out too soon. After a short flutter and glide, they end up clinging to a wall or landing on the ground. I have heard of people trying to help these birds by

attempting to launch them, but to no purpose. The instinct to fly was not yet fully developed. For such an aerial bird as a House Martin to be grounded sounds as serious as a fish out of water. These young martins ought to be doomed but they cope surprisingly well. They are more agile on their feet than one would expect and they have the instinct to hide under cover. Their parents continue to care for them, bringing food and encouraging them into the air with low passes. Eventually, something clicks and off the youngsters go, flying strongly and competently.

while they learn to forage for themselves. As their feathers grow and they become more practised, the fledglings are able to follow their parents. The advantage for the parents is that, with the family in tow, they can travel farther afield in search of food, beyond the confines of the old territory, and do not have to fly back to the nest to deliver each beakful.

Young birds leave before they can fly well because the nest is a dangerous place, not just because they are growing out of it but because they are magnets for predators. They leave as soon as is practically possible, which means before they can fly properly. A nestful of baby birds can be destroyed in a moment by a Magpie, Weasel or

squirrel and all the parents' efforts will be utterly wasted but, if the family is spread out, there is a good chance that some of the brood will escape a predator's attentions. Disturbance of the nest may result in nearly-fledged young jumping out prematurely, something that gardeners should be aware of. It is very difficult to round the birds up and stuff them all back in the nest because they keep leaping out again.

Family divisions

Some birds, such as the Great Spotted Woodpecker, divide the family between the parents. Blackbirds and Robins feed only the fledg-

parents. Blackbirds and Robins feed only the fledglings for which they have accepted responsibility (in Robins this seems to be young of the opposite sex) and reject the remainder of the brood. In their turn, the fledglings learn to beg for food only from their 'own' parent. Splitting the brood is most likely to occur when food is scarce and it could be a mechanism for preventing the entire family being eliminated in a single calamity.

Fending for themselves

After leaving the nest, the fledglings learn to feed themselves. They start by pecking at a variety of objects and gradually learn what is edible. Trial and error learning continues throughout their lives as they encounter new types of food in different seasons. They probably also learn by watching their parents. Every summer I watch a Great Spotted Woodpecker bringing a fledgling (identified by the red crown of its head) to the peanut feeder. Life is now much easier for the adult because, with the juvenile settled almost within reach of the feeder, it does not have to ferry nuts to the nest 100 metres (328 ft) away. It seems inconceivable that the youngster is not learning about peanuts and how to obtain them, but I have no way of proving this.

The break-up of the family comes gradually. The parents become increasingly reluctant to provide free meals and, instead of seeking out their young and offering food, they feed them only when they beg. At the same time, the young birds increasingly learn to fend for themselves and, when finding their own food becomes more profitable than cadging from unwilling parents, the family bond is broken. This is a neat mechanism for efficiently launching the young birds into independent life. If food is short, hungry juveniles are supported by their parents but, if plentiful, family ties are quickly severed.

ABOVE: Juvenile Linnets wait for a parent carrying food. As they learn to fend for themselves the family breaks up and the parents lose interest.

BELOW: Young Tawny Owls are entirely dependent on the parents for food for their first three months out of the nest.

ABOVE: When Tawny Owls leave their parents' territory, their survival will depend on finding a place to set up their own territories.

Exceptions to the rule

Surprisingly, young Tawny Owls do not hunt during their three months in the parental territory. They are totally dependent on their parents and spend much of their time calling with monotonous squeaking cries, telling their parents where they are and that they are still hungry. Some time is spent exploring the territory and it may be that the owls are closely observing the habits of mice, earthworms and beetles in their dark woodland home so that they will know how to locate them when they become independent. This is speculation but I cannot think of any other reason for such a long period of dependence. Juvenile Barn Owls remain together and are fed by their parents for no more than one month.

There are two other special, and contrasting, cases of upbringing. The Swift has no dependent period whatsoever and the fledglings fend for themselves as soon as they leave the nest. Long-tailed Tits, on the other hand, remain together as a family unit through the winter until breeding starts in the following spring. Families may join up and flocks of three or four dozen, perhaps even 60 or so, Long-tailed Tits stream across the garden like a flight of tiny arrows. For most garden birds, the fledglings are dependent for one to three weeks.

FROM FLEDGLING TO INDEPENDENCE

Species	Time (in weeks) to independence
Sparrowhawk	3–4
Collared Dove	1
Tawny Owl	12
Swift	0
Great Spotted Woodpecker	1
Dunnock	2
Robin	3
Blackbird	3
Great Tit	3
House Sparrow	2
Chaffinch	2–3

Life as an adult

Eventually the families of birds that were reared in the garden break up and the juveniles, recognizable in their drab plumage, gradually disappear. They drift away or are driven away in search of homes of their own or they moult out of their first plumage and become indistinguishable from their parents. I watched a Bullfinch rejecting its family when they were feasting on the dandelions and self heal that had flowered and set seed in my lawn. Once in a while, a fledgling, identified by the absence of black cap, flew to an adult to beg for food, quite unnecessarily in view of its ability to feed itself. The adult would take-off and settle a few yards away. The young bird was hoping for a free feed, but the parent was pointing out in no uncertain terms that its duties had finished.

Fostering

A correspondent told me of a Robin that fed two fledgling Wrens. This sort of behaviour is not unknown, but is unusual. The fledglings had crash-landed, one into a bush and the other to the ground. Immediately, a Robin perched on a fence nearby became agitated, flew down to the Wrens and back to the fence. Both Wrens then flew to the Robin and, in turn, begged and were fed.

Another account describes how birds of one species came to adopt a family of another species. A pair of Robins were nesting inside a garage, to which they gained access through a small window, beside which a pair of Wrens were nesting in a nestbox. Sadly, the Robins' nestlings died, but the parents then started to feed the young Wrens. The adult Wrens accepted this benign intrusion, and it was not unusual to see a Robin and the Wren at the nestbox together. Such was the strength of the attachment that the Robins joined the Wrens in calling the youngsters out of the nest. With the strong instinct for parent birds to stuff food into the brightly coloured, gaping mouths of their offspring, it is easy to see how, when the Robins lost their own young, the instinct could be transferred to the Wrens begging visibly at the entrance of the nestbox.

ABOVE: Young Cuckoos sometimes come into gardens with their foster parents. Other birds also benefit from fostering.

The feeding urge

I also have records of Robins adopting Spotted Flycatchers and Song Thrushes, Wrens adopting Blue Tits, and Blue and Great Tits adopting Wrens. It may be that that the fledglings become separated from their parents and are found by an adult of another species that has lost its own brood but still retains the instinct to drop food into gaping mouths. But there must be more to the behaviour than a parent bird, with its instinct to feed nestlings, reacting to the sight of a coloured mouth. Some birds seem to go out of their way to find young birds to feed. The urge to feed begging youngsters has caused a Wren to enter a nestbox to feed Blue Tit nestlings which it would not have seen from outside. Stranger still is the story of fledgling Wrens entering a Willow Warbler's nest and the parent warblers feeding both broods. The most ludicrous situation, however, arose when a young Cuckoo was enjoying the undivided attentions of a pair of Dunnocks, having thrown out their eggs, until it was joined in their nest by a brood of Wrens, who sheltered under the wings of the half-fledged Cuckoo.

A helping hand

A flock of Long-tailed Tits consists of two adults, their offspring and some additional adults. These adults are a remarkable feature of Long-tailed Tit life because they help rear the family in a sort of 'au pair' role, except that they are related to their charges. They are, in fact, paternal uncles and aunts which have lost their own families. Most nesting pairs have helpers, and there is usually one helper per nest but there can be as many as eight. The helpers join in when the eggs hatch and play their part in feeding the nestlings – a great assistance to the parents. The nestlings receive more food so they grow bigger and are given a good start in life.

PART 4

How Birds Survive

Life is not easy for garden birds even when we provide them with extra food and shelter. Despite all the efforts they make to survive, their lives are short and most small birds will not live to see their second winter. A roost on the sheltered side of a tree, a bath that softens the feathers to help preening and a fraction of a second delay responding to attack by a predator, all are small but vital factors that may make the difference between life and death.

*ABOVE: A House Sparrow sunbathes by raising feathers to let the heat of the sun's rays penetrate its plumage. **RIGHT:** The small size of the Wren makes it specially vulnerable to cold weather but communal roosting helps it survive long nights.*

Birds at night

Ghost-grey that fall of night,
  Ice-bound the lane,
Lone in the dying light
  Flits he again;
Lurking where shadows steal,
Perched in his coat of blood

WALTER DE LA MARE A ROBIN

Birds, like humans, are daytime animals and the great majority go to roost at nightfall. Although they are inactive during the hours of darkness, the way that they spend half their lives is important. A few species remain active after dark and the owls are the only species coming into the garden that are truly nocturnal.

There was a time when the drive home took me along a rough track running between grassy banks. As I approached in the gathering gloom a small bird would fly up in headlights. It was a Robin, like that seen by Walter de la Mare, which had been feeding in the last of the light. Robins have large eyes – this is what makes them so appealing on Christmas cards – and good night vision. Consequently they are one of the last birds to roost. Other birds retire to roost much earlier.

BELOW: Tits sometimes use lamps as roosts. The heat from the bulb ensures they will be kept warm, and so save energy, on the coldest nights.

Roosting

To see birds enter the roost and settle for the night requires patient observation or a little luck. Daytime reconnaissance will reveal tell-tale accumulations of white droppings under regular roosts. Then an evening vigil may be rewarded with a glimpse of the bird going into the roost.

Some, like the tits, slip away unnoticed but small flocks of Chaffinches and Greenfinches circle and alight in treetops before dropping into the roost. Although they do not pack tightly onto perches like Starlings on a city building, they gather into loose assemblies that are considered as communal roosts. Similarly, Blackbirds gather from bare, shelterless gardens to share a favourable roosting place in a shrubbery or hedge, to the annoyance of the local birds, which produce the chorus of metallic 'chink' calls that is so familiar on winter evenings. In select places several hundred Blackbirds may gather in a single roost.

House sparrows at home

I used to have a House Sparrow winter roost in the garden of a former house that was easy to watch. It was in a line of cypress trees and the sparrows started to arrive from all directions while it was still light and spend an age settling down. They first stopped in nearby trees and called to each other, then one would fly to the tip of a cypress branch, look this

ABOVE: Jackdaws fly towards their roost in a clump of trees. In winter they may not enter the roost until it is dark and mated pairs perch side by side.

way and that, hop to other branches and finally slip into the depths of the greenery. Others followed in turn and for a while there would be a chorus of chirruping that died down as the light faded. These sparrows were probably young birds because the breeders spend the winter roosting in their old nests.

Roosting ritual

Starlings are surprising because they start gathering in preparation for roosting so early in the day. Each afternoon in the winter months, while it is still bright, a small flock of Starlings settles in the top of a tree at the bottom of my garden. This is the start of a ritual that I consider to be one of the greatest spectacles of everyday wildlife. Whether in town or country, Starlings gather from miles around to spend the night in communal roosts that

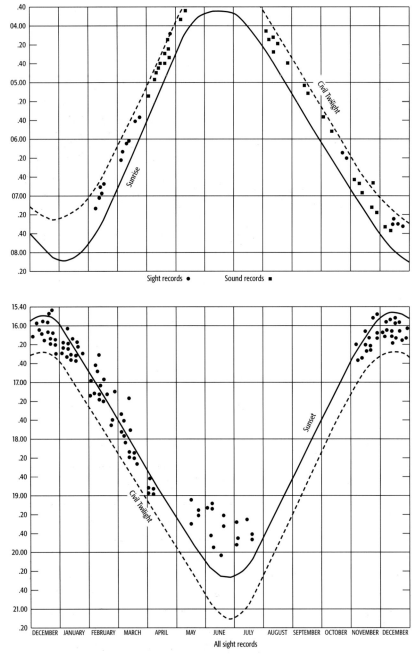

Sight records ●   Sound records ■

All sight records

They are seen best on a clear evening when the flock is flying against a sky glowing orange or pink in the setting sun. Several minutes may pass as the flock stretches out into a wisp, gathers again into a compact bunch, drops earthward and rears back up, roller-coaster fashion. Smaller groups detach themselves, land, take-off and land again until, eventually, all are safely perched for the night. Although roosts are used throughout the year, they are much smaller in summer because they comprise only non-breeders and males off-duty from nesting. In winter, the whole Starling population gathers into these roosts and numbers are swollen further by immigrants from continental Europe.

Twilight activity

Other birds indulge in fewer preliminaries and their entry into the roost and departure in the morning are closely linked with the time of dusk and dawn or, more precisely, to civil twilight. (This is the time at which the sun is six degrees below the horizon and is between 30 and 60 minutes after sunset or before sunrise. It is roughly equivalent to lighting-up time.) When breeding, birds, not surprisingly, rise earlier and retire later. Street lights and floodlighting often keep birds awake and Robins especially can be heard singing at night. However, loud noises, like gunfire and thunder, may also stimulate a nocturnal chorus.

ABOVE: These two charts show the times at which Wrens awake (top) and roost (bottom) at different times of the year.

may number hundreds of thousands of birds. My flock joins other Starlings flying over and increasing masses of birds converge on their roost in a wood on the skyline.

The Starlings do not enter the roost at once, but fly to and fro in swirling clouds of packed birds.

Favourite places to roost

From my window I also see flocks of gulls, Woodpigeons and Rooks streaming over to their communal roosts in the late afternoon. Other kinds of birds are usually more solitary and certainly more elusive as they enter their roosts. Favourite places for roosts include ivy and cotoneaster growing up walls, and the evergreen foliage of conifers, including hedges of the reviled leylandii. Small birds also use natural crevices and abandoned nests or man-made structures. Robins usually roost in foliage but, as the poet suggests, they may use a barn or shed.

Wrens like the cover of nestboxes, tree holes and buildings. In severe weather, adults roost together and keep each other warm. The movement seems to be started by the cock in whose territory the roost is situated. He sings a rallying call to other Wrens, which are drawn in from up to a mile or more. There are usually fewer than 10 birds in a roost, but there is a record of 61 packing into one nestbox. The Wrens arrange themselves in neat tiers, one layer perched on the backs of those below, heads inwards and tails outwards. Wrens are unusual because a brood of juveniles roosts together after they have left the nest. One parent, usually the father, leads the fledglings into a nursery roost which is most often one of the unused 'cock nests' that he had constructed as part of his courtship routine (see p89).

Safe roosting

A roost located in thick vegetation offers some protection from predators, although Tawny Owls sometimes hunt roosting birds by diving into the foliage and snatching a victim off its perch. They also drag adults and young from nests under the cover of darkness. If Tawny Owls were active by day, they would be as unpopular as Magpies! However, roosting sites are chosen primarily for the refuge they give from the elements. Even a tangle of bare twigs gives significant shelter from a biting

ABOVE: *Mallards roost in the open where it is hard for predators to approach unnoticed. There are always some ducks awake and on guard.*

MURMURATION

Starlings wake at dawn and start to chatter among themselves. The sky becomes brighter and suddenly there is a lull in the babble as if the switch has been thrown. A few seconds later, the murmuration, as the Starlings' chorus of chattering and whistling is called, starts again. Bursts of sound alternate with silence, then, just as suddenly, a mass of Starlings flies out of the roost in a dense cloud. A few minutes later, there is another hush, and another exodus of birds. Murmuration and departures recur until the roost is empty. Waves of Starlings spread out from the roost over a radius of 15 to 20 miles (24–32 km). In the reverse of the process in which the flocks gather to go into the roost, they split up on their way out as small groups, drop out to feeding places, while the main mass passes on.

TREECREEPER ROOST

Whenever I come across a large wellingtonia tree with its characteristic thick, spongy bark which you can punch without hurting your knuckles, I examine the trunk closely. I look for deep, round pits in the soft, spongy bark, about 5 cm (2 in) across and 3 cm (¾ in) deep. These are the roosting holes of Treecreepers. Treecreepers also excavate similar cavities in rotten wood but I have never found them. The pits are usually about head height, on the sheltered side of the trunk, and a dribble of white droppings under a pit is a sign of recent occupation. The Treecreepers use their pits on cold nights and fly in at sunset. Sometimes several Treecreepers huddle in a single pit, heads inward and tails radiating like the spokes of a wheel.

ABOVE: *A flock of Long-tailed Tits roosts on a perch. They huddle more closely on cold nights and dominant birds get the middle, warmer places.*

wind, and a Blackbird roosting in cover saves as much as one third of the energy needed to keep warm on an exposed perch. Rooks even compete to secure places on the leeward branches of a bare tree, which offer a small but significantly greater amount of protection over the windward side. Birds that creep into cavities gain a much greater degree of protection.

Warm roosts

Tits prefer to make their winter roosts in nestboxes or natural tree holes where they are better protected from the cold and predators. If none are available, they have to roost in the open. Sometimes they find the perfect solution and take up residence in outdoor lights, where they are not only safe but very warm. One Blue Tit roosted in a light over a garage door. Every night it started on the bulb and dropped to the floor of the lamp when it got too warm.

Nocturnal birds

It is a shame that the Nightingale cannot be considered a garden bird although it may come into large gardens with dense shrubberies. It can be a mixed blessing because I have heard of a family who moved house because the nocturnal outpourings of a Nightingale upset their sleep!

Otherwise, the garden's nocturnal birds are limited to Tawny Owls, and perhaps Little Owls. Some people are upset by Tawny Owls hooting close to the house but for most of us this is a wonderful sound, especially when the silence of the night is broken by one owl hooting nearby and answered by another in the distance. They seem to emphasize how empty is the world at night. They do not, of course, call 'tu-whit tu-woo'. That was Shakespeare's idea. Both sexes call with metallic 'kewicks' and the long drawn-out 'oo-hoo-ooo' but the one call followed by the other will be a female followed by the male.

Tawny Owls

Young Tawny Owls stay with their parents for three months after they have left the nest and they can be heard calling with a shrill, monotonous 'ker-sip' through warm summer nights. Hunting sometimes takes place by day in summer when nights are short and there are hungry mouths to fill but most hunting is nocturnal, when small mammals, earthworms and beetles are active. However, urban Tawny Owls, especially, hunt roosting and nesting birds up to the size of Woodpigeons. Although they have good vision, acute hearing is the main sense for pinpointing prey in pitch darkness. (Surprisingly, their senses of

ABOVE: *Although generally nocturnal, Little Owls can sometimes be seen perching on branches or posts, and occasionally hunting for worms in compost heaps.*

vision and hearing are not much better than our own.) They can pick up the rustling of a mouse under snow but hunting is hampered by the pattering of raindrops drowning the sounds of prey.

Sleep

The North Wind will blow,
And we shall have Snow,
And what will the Robin do then, poor thing?
He'll hide in the Barn to keep himself warm,
And hide his Head under his Wing, poor thing.

ANON

Anon was wrong because birds, with the exception of penguins, do not put their heads under their wings. They turn their heads and thrust their bills under the scapulars – the feathers covering the 'shoulders' - and into the warm air trapped in the feathers. It is the equivalent of pulling the eiderdown over your head. Fluffing the feathers traps extra air to increase the insulation. Alternatively, when sleeping lightly, the bill is laid along the back or the bird simply withdraws its head into its shoulders, looking straight ahead.

The function of sleep is not properly understood even in humans. It seems that it is a way of 'switching off' when the animal has nothing else to do. It saves energy and its inactivity may also render it less conspicuous to predators.

Flying when asleep

Birds do more than sleep in the roost. They preen, interact with each other and occasionally sing. Neither is sleeping restricted to the roost. They sleep at intervals during the day when they are inactive or 'loafing'. Even when sleeping deeply, a bird opens one or both eyes, or 'peeks', at frequent intervals to look out for danger. So a sleeping bird is difficult to approach. If it has been disturbed or is roosting in an exposed place peeking is more frequent but the bird peeks less when roosting in a group and guard duty is shared.

Although disputed by some

LEFT: *Pied Wagtails gather to roost, sometimes in well-lit, noisy town centres. They do not huddle for warmth but the mass of birds is safe from surprise attack.*

ornithologists, Swifts are believed to roost and sleep on the wing. It is a common observation that flocks of Swifts circle high into the sky at dusk until lost from view. It is also known that they can be seen descending early in the morning. In World War I, a French pilot gliding over the lines by moonlight, with engine cut, passed through a flock of birds at 3,000 metres (10,000 ft). Next day, a Swift was found caught in the plane. There is more circumstantial, less definite evidence that House Martins may spend the night in the air (see p 126).

ABOVE: *A Greenfinch asleep on the ground.*
BELOW: *Blue Tits use their nesting places as snug, safe roosts in winter.*

Care and maintenance

She plumes her feathers, lets grow her wings
That in the various bustle of resort
Were all to ruffl'd, and sometimes impair'd

JOHN MILTON COMUS

A bird's feathers are a unique covering. They are vital for keeping warm and flying. If they are not kept in good condition, the bird's health and well-being are threatened on both accounts. Feathers, like finger nails which are made of the same material, are dead structures. Unlike nails, they do not grow continuously and they wear out. They cannot be repaired if damaged and replacement of worn and damaged feathers has to wait until the annual moult in late summer. As the integrity of the plumage is so important it is not surprising that birds devote a good deal of time to its care and maintenance.

Preening

This is the main form of feather care and takes place at odd moments through the bird's waking hours. It may consist of a few seconds' rearrangement of some feathers, perhaps if they are uncomfortably out of place, but serious preening sessions are conducted when the bird is loafing and not engaged in any other specific activity.

Birds preen by gently nibbling their feathers as they are drawn through the bill. This action cleans away dust and debris and 'zips up' splits in the vane. A second but important function is to remove parasites, such as blood-sucking fleas and ticks, and feather lice which feed on the bird's plumage. Songbirds with short bills modify the nibbling action into pecking rapidly at the feather with the tip of the bill, as if chipping rust. A second action consists of stroking the feathers with the edge of the bill, or sometimes the head, to smooth them into place. During serious preening sessions, the bird twists its body and squeezes out oil from the preen gland under the tail. It is believed that the oil helps waterproof the feathers and preserve their structure, perhaps through having antibiotic or fungicidal properties.

LEFT: During preening, a Jay runs a tail feather through its bill to 'zip up' and clean the vane.

Bathing

Many bird books ignore or barely mention the subject of bathing although it can be seen in every garden that has a birdbath, pond or puddle. The subject is, however, not so straightforward as it might seem.

It is sometimes said that birds bathe to keep their feathers clean, and even that city birds have a greater need to bathe than their country cousins. There is no evidence that birds try to keep clean in the same way as we do, although they make an effort to remove severe fouling of the feathers.

Popular bathing areas

I once wrote in a magazine article that I hardly ever saw birds making use of my birdbath, even though it was outside my study window and so kept under almost continuous observation in daylight hours. I have never received so much correspondence! In 150 letters, the key word was 'surprise': surprise that birds were not using my birdbath. The problem was probably that my birdbath was in the wrong place. I later noticed birds bathing in the gutter on my garage and fixed a bath to a bracket on the wall. It

quickly became a great attraction. As with nestboxes, it is not altogether clear why one birdbath should be popular and another shunned. It may have something to do with security. A bird with wet plumage is hampered if it has to escape from a predator and the birdbath on my garage was close to an evergreen tree where bathers could quickly take refuge.

Bathing technique

Bathing is either a brief 'in and out' when the bird stands on the edge of a birdbath or puddle and hops in and out, like a human bather lacking courage for a

BELOW: A Mallard drake washes vigorously by alternately dipping its body into the water and rising to flap its wings. Washing is as necessary to waterbirds as to land birds.

ABOVE: *Swallows, like some other aerial birds, bathe by dipping into the surface while airborne and then shaking the excess water from their plumage.*

BELOW: *A House Sparrow's thorough bathe leaves it looking very bedraggled. The function of bathing is not completely clear.*

full immersion, or it is a good soak to let the water percolate through the feathers until the bird looks bedraggled. The absence of standing water does not prevent birds bathing. I have been entertained to the sight of a Swallow bathing in the swimming pool of our holiday home. It flew low over the pool several times, and made 'flying belly-flops' into the water, checking momentarily as spray billowed around it and shaking itself as it flew on. Swallows have also been seen going through the same motions on dew-soaked lawns and taking quick 'showers' by flying through spray from waterfalls. Other birds also make use of damp lawns and foliage or have 'shower baths' while perched in the rain.

The first bath

A young bird's first bath can be amusing. A fledgling Robin watched some House Sparrows in a small bird-bath soaking the surrounding grass with their splashing. After the sparrows had gone, the Robin hopped onto the rim of the birdbath but did not venture in. Instead, it dropped to the ground and fluffed out its feathers and rubbed its body and head in the wet grass. It seems this young Robin was stimulated to try a bath by the sight of the other birds. It had an idea of what to do, but was not quite sure how and where to do it. I have seen something similar when a Song Thrush brought its fledglings into the garden. As each juvenile noticed the pond, it jumped in and leaped out again as it found itself out of its depth. Each fledgling then went through the motions of bathing on dry land beside the pond.

The function of bathing

Bathing may clean the plumage but it is more likely that the main purpose is to damp the feathers to facilitate the spread of preen oil over the feathers and soften them so they become easier to work into shape and position. Maintenance of the plumage would explain why birds bathe on cold days (although if it is too cold problems can arise – a Starling was once seen to become flightless because its plumage had iced up). Plumage in top condition will give birds the most effective insulation from the cold air.

It seems to me, however, that the function of bathing by birds is still something of a puzzle. From my own observations, I get the impression that while Blue Tits and Blackbirds bathe regularly, Chaffinches and Greenfinches rarely bathe and I have never seen the garden's regular Great Spotted and Green Woodpeckers or Dunnocks bathe (although these species are known to bathe). If bathing is essential for keeping the plumage in good condition, I would expect to see more birds making use of my birdbaths. The Rev. E. A. Armstrong, who wrote a monograph on the Wren based on more than 10 years close observation, only once saw a Wren bathing, although he does seem to have been particularly unlucky. The key question is how frequently individual birds bathe. It would take an enormous amount of time and effort to ascertain this fact, but it might shed light on the function of bathing.

ABOVE: *The Pheasant is one of the species that regularly dustbathes. The dry soil is thought to help remove the impurities from the feathers.*

Dustbathing

If bathing in water is to clean the plumage, dustbathing must be the antithesis. The action is very similar. The bird squats on bare soil in a sunny place, perhaps in a flower bed or a dusty path, and makes the same movements as in bathing to flick dust over its body. Afterwards it shakes itself vigorously and indulges in a bout of preening. The function of dustbathing is believed to be to remove excess water, preen oil or parasites.

Sunbathing

It is not unusual in the summer months to see a bird sunbathing on the lawn or the top of a fence. Sunning is a better term than sunbathing because, as in humans, the functions of sunbathing and bathing are different.

When sunning, the bird raises the feathers of its head and body to let the sun's warmth penetrate and usually spreads its wings and tail. It often looks as if it is in a trance with staring eyes and bill agape. There are two types of sunning, both having functions more presumed than proven:

Sun-exposure

This is seen in the heat of the day and several possible functions have been suggested. There is some evidence from experiments that feathers that have been bent out of shape will be more rapidly restored if warmed in the sun. Another possibility is that sunlight may convert preen oil spread over the feathers into vitamin A, which is later ingested when the bird preens. A third idea is that the sun's warmth will make parasites more mobile and easier to detect and catch.

ABOVE: *Flocks of House Martins land on roofs and expose themselves to the rays of the sun.*

Sun-basking

This is seen in cool weather and seems to be a way of warming up. A friend used to live in a fifth floor flat from which he could see hundreds of House Martins clustered on the roof of a nearby building in the early morning. The roof faced the rising sun and the black slates absorbed the warmth of its weak rays. The martins stood with their backs to the sun and sunned with their wings spread. It is believed that House Martins roost on the wing overnight, as Swifts are known to do. Birds which have been caught after descending in the morning are cold to the touch. Sunbathing on a warm roof would be as beneficial as huddling round the fire after a cold walk is for us.

ABOVE: A Blackbird suns itself by spreading its wings and tail, and raising the feathers of the back to allow the sun's rays to penetrate.

BELOW: This Magpie has chosen a secluded but sunny corner of the garden to sunbathe. It looks as if it is in a trance but it will fly away at the slightest disturbance.

ANTBATHING

Antbathing or anting is a strange behaviour that is something of a mystery. Although recorded in around 200 species of songbirds, it is not often seen but is eye-catching when it is. A bird on the lawn, perhaps a Starling or Song Thrush, spread-eagles itself on the ground, rather as if it is sunning, and with the nictitating membrane (so-called third eyelid) covering the eye. Alternatively, a bird spreads its wings and twists its tail, throwing itself into contortions as if having a fit.

The common feature is that these birds are on an ants' nest. The first bird is simply allowing the ants to crawl among its plumage. The second picks them up and rubs them into the wing and tail feathers. It has been suggested that formic acid or other secretions from the ants' bodies have fungicidal or insecticidal properties, or help to remove stale preen oil. Yet, as with sunning, there appears to be something about the birds' excited manner that suggests that anting is more than strictly functional and may be pleasurable.

Moulting

Over the course of the year, a bird's feathers become worn. Pushing through foliage, constant rubbing against the nest bowl during incubation, the ravages of feather mites and fungi and even the stresses of flying, all contribute to the gradual abrasion of the feather vanes.

By the time the family is launched into independence in late summer, the parents' plumage has faded and feathers are beginning to disintegrate. It is time for the annual moult in which most, if not all, the bird's feathers are shed and replaced by new ones. As each new feather grows, it pushes the old one out of its socket, but there is a gap in the plumage until the new feather is full-grown.

A slow process

The moult is slow and takes about 12 weeks in thrushes and finches and 16 weeks in tits, depending on age and species. Feather replacement is an

BELOW: Two young Starlings moulting out of their dull, grey-brown juvenile plumage into the glossy, spangled adult winter plumage.

orderly process and feathers are lost symmetrically from wings and tail. It goes unnoticed unless the bird can be examined in the hand. The exception is that you can see the gaps where flight feathers have been shed from the wings of Rooks or crows as they glide overhead.

Many risks

During the moult the bird needs extra energy to grow new feathers, while at the same time, flying and keeping warm are less efficient because of the gaps in the plumage. In this state the bird will be more at risk from predators. The situation is made yet worse because the moult takes place after the nesting season when birds may be already in poor condition from the effort of rearing their families. Sometimes they are in such a feeble state that feather replacement has to slow down. Young birds are even worse off and may die if there is not enough food to support the moult. The strain of the moult is one reason for the common complaint that birds have deserted the garden. While in this vulnerable state, they become less active. They avoid long flights and spend their time resting under cover.

Bald birds

It is not unusual to see birds with heads and necks completely devoid of feathers during the moulting season. In my garden there was a Magpie whose bald head made it look like a ridiculous caricature of a tiny vulture. Baldness is a consequence of something going wrong with

ABOVE: The striking contrast between a male Brambling in summer plumage (right) and winter plumage (left) is due to the tips of the feathers wearing away to reveal bolder colours underneath.

RIGHT: Gaps in the wings of this Carrion Crow show where the flight feathers have been shed. Half-grown feathers can be seen in the gaps near the wingtips.

the normally orderly sequence of feather replacement. The bird goes bald because the head is the last part of the body to moult and, under extreme conditions, the feathers are shed before the new feathers are ready to grow. A bald head creates an even greater stress because the brain is the one part of the body that must be kept warm at all costs. So everything is conspiring against these odd-looking birds and they should be objects of pity rather than ridicule.

Survival

Assuming that each robin lives for ten years and that each pair produces ten eggs a year, then, if they and their offspring are unmolested, a single pair will in ten years have multiplied to more than twenty millions. As there is no reason to think that the robin population is increasing at all, it is clear that an exceedingly large number of robins must die or be killed each year. We see nothing, or almost nothing, of this tremendous slaughter of innocents going on all around us.

ALFRED RUSSEL WALLACE DARWINISM 1889

Many birds living in the garden at the end of summer will be dead within a year. Most of the eggs laid by the survivors will not result in young birds becoming independent. There are many dangers facing nests and adult birds, ranging from spells of bad weather and accidents to disease and predation. We are unaware of this struggle for existence except when we witness the high-profile actions of predators.

BELOW: The attentions of a Cuckoo are one more problem for small birds trying to rear a family despite the hazards of food shortage, bad weather and predators.

A short existence

In the western world, we expect that almost every human baby will survive infancy and nearly every child will grow up, lead a long adult life and eventually die in old age. It can come as a shock to find that, in Nature, the opposite is true.

It may not be obvious that the overwhelming majority of animals die young. Alfred Russel Wallace's simple arithmetic shows that if there were not a huge mortality of Robins, we would be knee-deep in them. Fifty years ago David Lack was able to demonstrate the true state of affairs in Robins by examining the records for ringed birds. Of the adult Robins alive at the end of the nesting season, 60 per cent, or over half, will be dead one year later. To put it another way, an adult Robin has the expectation of no more than a further 13 months of life. The more vulnerable juveniles have an even shorter expectancy. Proof that their lives have been cut short comes from the long lives of captive birds and rare evidence of long-lived wild birds. There are records from bird ringing of a Blue Tit living to 10 years, a Robin to 13 and a Blackbird and a Starling to 20.

Dwindling numbers

Throughout the spring and summer, birds work hard to bring up their families and the garden is eventually filled with small birds. But from the time they leave the nest their numbers start to dwindle. In one study of garden Blackbirds, each pair produced an annual average of four fledglings that left the nest, but fewer than two of these survived to the following spring. This is more than adequate because less than one bird on average is needed to replace the deaths among adults. By the beginning of the nesting season, the number of survivors will be about the same as had started to nest in the previous spring.

To summarize: a garden bird can expect to live for a year and rear one offspring to replace it. There is, however, considerable variation. A Dutch study showed that, among Great Tits, half the females failed to produce a single young bird to survive them. The population was maintained because a successful minority reared several young.

ABOVE: *When snow covers the ground, it is hard for Wrens to find food and there is a huge mortality. Succeeding mild winters allow their numbers to build up again.*

BELOW: *The Hobby specialises in hunting birds in flight. Its late nesting allows it to feed its nestlings on the inexperienced young of small birds.*

Causes of death

Given the huge mortality of birds, the next surprise is that so few corpses are seen. Most bodies simply disappear because they are either eaten by predators and scavengers, or because they are most likely to die under cover and their frail bodies soon disintegrate.

If a bird's body is found, it is not easy to work out the circumstances that killed it. Even if it has been the victim of a predator, there is the possibility that its survival had already been put in jeopardy by starvation, disease, parasites or injury. And if the bird dies of starvation, this may have been compounded by dominant members of the species keeping it from scarce food supplies or a warm roost. Just to make things more complicated, a well-fed, fat bird, although healthy, is more likely to be caught by a hawk because it is less agile! There is also a slow attrition caused by death through accidents such as collision with windows, overhead wires or motor vehicles. As these occur around human settlement, they are more often reported. There are also some strange causes of death, like the Goldcrest that was trapped in a spider's web. Finally, in this day and age, poisoning can be a significant cause of death.

Starving to death

However, it seems that starvation is a major cause of death among garden birds. Nests fail when parents cannot find enough food for their brood. For tits this can happen when rain washes caterpillars off leaves. Blackbirds and Song Thrushes are in difficulty when drought drives worms and insects deep into the soil. Garden birds often fare worse than country birds because gardens are comparatively barren unless we provide extra supplies of food. Winter is seen as the time of shortage, and birds such as Wrens are especially at risk when snow and ice lock up food, but summer brings problems because both nesting and the moult that follows require plenty of energy, and prolonged rain or drought can leave birds facing starvation.

Killed by predators

It is difficult to tell how many nests are destroyed by predators. The percentage varies from place to place and year to year but it can be very high. There is evidence that at least some of the losses to predators would have happened anyway from other causes. For instance, the begging calls of hungry nestlings attract predators but persistent calling is a sign that the nestlings are hungry and in danger of death by starvation. Predation can be serious in gardens because nests may be not be well-hidden and there are large numbers of cats and Magpies on the prowl. The parents may also be at risk because they have to concentrate so

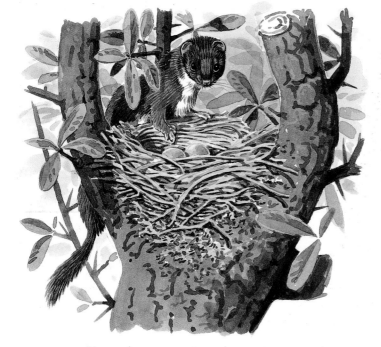

LEFT: *The Weasel is a good climber and an accomplished nest robber. It may be attracted by the cries of hungry nestlings.*

much on finding food that they become careless of the threat from predators.

Bad parenting

Nests can also fail because the parent birds are incompatible or inexperienced. This can lead to the eggs being infertile or the young being inadequately fed. Nestlings may be squashed or fall from the nest. Wind and heavy can rain destroy nests, especially when they are in exposed sites.

BELOW: *A Song Thrush's second clutch of the year is safer because it is hidden from predators by dense foliage.*

WINDOW STRIKES

With more picture windows and patio doors and more birds coming into gardens, the sad result is that literally millions of birds collide with windows every year. Some bounce off and fly away. Others are dazed or stunned. (Place them in a cardboard box and keep them dark and warm until they have recovered.) One third, it is estimated, are killed.

Some windows are more dangerous than others. They may be near a feeder or nestbox. A window at the other end of the room or the sky's reflection may give the impression of a through passage. Silhouettes of birds of prey or spiders' webs stuck on the pane act as a warning. Net curtains or non-reflective clingfilm stuck on the outside of the pane are other possible solutions if a window is found to be particularly dangerous. There are no complete solutions to the problem.

Sometimes the bird leaves a ghostly imprint of its outline. It may even show individual wing feathers. The force of the impact has shaken out microscopic particles of feather dust or powder down from its plumage. Some birds, such as pigeons and doves, produce large quantities of down and leave conspicuous imprints.

Eluding enemies

The most important enemies of adult garden birds are cats and Sparrowhawks. Kestrels and Tawny Owls prey mainly on small mammals, although owls sometimes hunt roosting birds, and Magpies, Jays, Weasels and squirrels are primarily nest-robbers. Both cats and Sparrowhawks hunt by surprise attack but the threats they pose and the response of their quarry differ because the former hunts on the ground and the latter in the air.

ABOVE: *Given a head start, a Great Tit has managed to reach the safety of a thick bush. The Sparrowhawk, however, is very good at following its victims with its long legs.*

BELOW: *This Sparrowhawk has caught a Blackbird. It was probably by a surprise attack that left the small bird no time to escape.*

The Sparrowhawk's technique is to appear out of nowhere, around a corner, through a gap in a fence or over a hedge, and catch small birds unawares. If you see a flurry of small birds disappearing from the birdtable you might also see a Sparrowhawk hurtling through the garden. The attack is not without danger to the Sparrowhawk because even a slight injury could jeopardize future hunting. I once saw a Sparrowhawk chase a sparrow under my parked car and bounce off the metal garage door beyond with a clang as the sparrow flew to safety.

An unusual sight

A Sparrowhawk would seem to have an overwhelming advantage over its small prey, but only once in a while do I see one catch a bird. Sparrowhawks and other predators may look like mean hunting machines that have the odds heavily stacked in their favour. But one study found that only one in 10 Sparrowhawk attacks is successful. They have to work hard for a meal! Nevertheless, Sparrowhawks are a major cause of death among small woodland and garden birds.

This does not mean that they have caused the recent decline in small bird populations, as is sometimes suggested. Professor Ian Newton, who studied Sparrowhawks and finches throughout his working life, calculated that, in one year, a pair of Sparrowhawks accounted for 2,200 House Sparrows or 600 Blackbirds or 110 Woodpigeons. In woodland, Sparrowhawks are responsible for the deaths of a third of all fledgling Great Tits. Yet neither Professor Newton nor any other ornithologist has been able to demonstrate that this carnage has any long-term effect on the populations of prey species. The exception is around the immediate vicinity of the predators' nests where the local population of tits is greatly reduced.

The Magpie problem

The Magpie must be the most hated bird in the country. It is also one of the most mis-understood. Its evil reputation for destroying nests and wiping out a neighbourhood's small birds arouses such passion that I risk my own reputation by discussing its habits.

An unfussy predator

The Magpie is an opportunistic omnivore that will make use of almost any source of food, from robbing milk bottles to hunting mice (which, ironically, once made Magpies popular with farmers). There is no doubt that Magpies frequently raid nests for eggs and nestlings, but, as with the Sparrowhawks' predation on small birds, it has proved impossible to show the scale of depredation has a significant effect on bird populations, except at a local scale. Adult birds such as sparrows are sometimes hunted but this is rare and does not affect populations, although it is sad to witness and a tragedy for the individual birds.

Songbirds survive

An analysis of 35 years of records by the British Trust for Ornithology has failed to show a link between the decline of our common songbirds and the increase in Magpies. There is no difference in the numbers of songbirds in places where Magpies are numerous and where they are scarce. This is no con-solation to anyone who has watched a Magpie perched high in a tree overlooking their garden and moni-toring the movements of the small birds. And then witness it drop like a stone into a bush and emerge with an egg or nestling. It seems that some Magpies systematically patrol their territories to rob nests. Yet even where nest losses from the depreda-tions of Magpies and other mem-bers of the crow family, squirrels, cats and small boys are high, birds still manage to rear enough nestlings to maintain the population. The excep-tion seems to be in barren places like urban parks, but poor breeding there is balanced by immigration of surplus birds from more favourable habitats.

A man-made problem

The Magpie problem is largely man-made. Magpies have increased in the countryside because there is less persecution by gamekeepers. They have increased even more in towns because extra food from birdtables, discarded take-aways, bin-bags and road-deaths gives them an abundance of food for winter survival. They even get a measure of protection from their own predators because, by nesting close to buildings, they are safe from Carrion Crows, themselves nest-robbers but more wary of human presence.

BELOW: *The sight of a Magpie carrying off a helpless nestling does nothing to enhance its bad reputation.*

Keeping alert

Whatever the threat, a bird's first line of defence is to keep alert. Watch any bird when feeding: it is continually breaking off to look around. Reacting in a split second can make the difference between life and death, but pausing to look around wastes feeding time.

Birds that feed in flocks have an advantage because many pairs of eyes make it harder for a predator to pounce unnoticed. In the same way as feeding in a flock is an advantage because many eyes are better for spotting food (see p 75), if one bird spots approaching danger, its alarm call or a rapid escape alert the rest. So individual birds in a flock are able to spend more time feeding and less time watching for danger because they are covering for each other.

An eye out for danger

Observations on Starlings show that a bird feeding alone is vigilant for half its time but, in a flock of 10, each bird spends no more than about one tenth of its time looking around. Moreover, with many pairs of eyes on the look-out at any moment, the overall level of vigilance is higher in the flock and its members react to danger significantly faster than a bird on its own.

The safety of numbers

Another advantage of living in a flock is that each bird is statistically less likely to be the unlucky one when the predator pounces. This is the selfish concept of safety in numbers. A solitary bird is in great danger if a hawk swoops on it, but if it is in a flock one of its companions may be the unlucky one. In a flock of 10, each bird has only a 1 in 10 chance of being caught; if there are 20 birds, it is even safer with a lower, 1 in 20, chance of falling victim.

ABOVE: The Great Spotted Woodpecker is an unexpected nest-robber but it either reaches through the entrance or drills through the walls and extracts the nestlings. LEFT: A Grey Squirrel poses a similar threat. A metal guard stops predators enlarging the entrance hole of a nestbox.

One intriguing discovery made by ornithologists studying sparrows at a feeder shows that they recognize the safety of numbers. When a sparrow spots food on the ground it often lands on a nearby perch and starts chirruping. This attracts other sparrows and they all descend to feed. On the face of it, this is rather silly: the first sparrow ought to keep the food for itself, not share it with its fellows. However, a solitary sparrow feeding on the ground is, as we have seen, very vulnerable, so it is worth sharing its food for the extra safety its companions confer. On the other hand, when opportunities for feeding are limited, the sparrow stays quiet and keeps the food for itself, despite the extra risk.

When attacked in the air by a Sparrowhawk, a flock of Starlings bunches together. If you see a flock circling in tight formation, look for the Sparrowhawk, or perhaps a Kestrel. The mass of birds confuses the hawk and makes it difficult to single out a victim. The hawk also has to take care because the packed mass of birds increases risk of a damaging collision.

The scatter effect

The alternative to bunching together is to scatter. I saw a good illustration of this when a large helicopter flew over the house at about 100 metres (109

ABOVE: *Blackcaps and other small birds survive by remaining alert. They are continually scanning their surroundings for danger and also listening for alarm calls from other birds.*

yards). As it disappeared, the sky was filled with a swarm of Woodpigeons. They were spreading in all directions, darting hither and thither with every sign of panic, but their response was appropriate for dealing with an attack by an aerial predator like a Peregrine. If a pigeon cannot reach the safety of cover, its best chance is to outmanoeuvre its pursuer, and also spoil its aim by jinking erratically. The same tactic can be seen on the rugby field when a running player feints to left or right to wrongfoot a tackler.

ALARM CALLS

Birds' alarm calls are often heard in the garden if only because they are directed at us! There are two main kinds: alarms directed at danger on the ground and alarms directed at aerial predators. If you are working in the garden you may become aware of repeated sharp 'ticking' or churring notes coming from a hedge or bush. It is probably a sign that you are too close to someone's nest. Or else a cat is. These calls give warning to other birds of the approach of a mammalian predator. Providing the birds know where the danger lies, and remain alert, they will be safe.

A swooping Sparrowhawk is a different proposition. A bird that spots its approach gives a thin, high-pitched whistle – *seeeee*. The caller runs a risk by giving away its location but the physical characteristics of this sound make it difficult for the hawk to pinpoint the position of the caller. On hearing the alarm, all the small birds within earshot rush for shelter. If they are already under cover, they freeze until the danger is past.

Mobbing

The best form of defence is said to be to attack and it is not unknown for small birds to turn on their attackers and send them packing. This is usually the case when they are defending their nests.

Mistle Thrushes and Swallows are famous for harassing predators in the vicinity of their nests. There was an occasion when I witnessed dirty work high in a horse chestnut where a Carrion Crow was molesting the nest of a Mistle Thrush and the parents were trying to defend it. As a species, the Mistle Thrush has a reputation for courage in defence of its young, but the efforts of this pair, while running true to form, had little effect on the predator. The crow was lifting a well-grown nestling

ABOVE: Small birds gather around a Tawny Owl. When it flies, they will follow it and their alarm calls alert other birds.

from the nest. The cries of the parents were answered by three more pairs of Mistle Thrushes. These converged on the crow, which was unhurriedly seeking the best grip on its victim, before flying off surrounded by the posse of eight aggressive and vociferous Mistle Thrushes.

Mob rule

It was rather strange that the other three pairs of Mistle Thrushes joined in the attempted rescue, because they were not protecting their own offspring, and so seemed to behave in a truly selfless and altruistic fashion. This is, however, more like the behaviour that can be seen when adult birds gather to mob a predator. If you hear a chorus of alarm calls and see a gathering of small birds in a tree, look for the villain. It will probably be a Tawny or Little Owl, a cat, a Sparrowhawk or Kestrel, perhaps a Jay. It may be on the prowl, although an owl may simply be roosting and minding its own business.

Risky business

The little birds may seem to be putting themselves at risk, especially when the predator takes off and they stream after it. But surprise is an essential element in the hunting behaviour of these predators. They must recognize that the game has been given away by the gang of small birds and that every other creature in earshot is aware of their presence. The potential victims realize this as well and show little concern when they can keep track of the predator. I have watched a Fox walk past a Pheasant and sit down, while the Pheasant continued on its way, slowly stepping past the Fox. There was no visible recognition of each other's presence but they must have been aware of their potential as predator and prey. It would be wrong to think that, because birds are always alert, they, and other animals, live in a perpetual state of fear. They treat Sparrowhawks or Foxes as we treat motor cars. They are potentially lethal but we learn to live with them by being safety conscious.

BELOW: Swallows are among the species that boldly attack predators much bigger than themselves. Even if the cat is not driven away, it will have no chance of successfully stalking a victim.

ABOVE: Starlings band together to mob a Sparrowhawk. The tight formation makes it difficult for the hawk to single out a victim.

Useful Addresses

The Wildlife Trusts
See Page 7 for information.
The Kiln, Waterside, Mather Road, Newark,
Nottinghamshire BG24 1WT
Tel: 0870 036 7711
www.wildlifetrusts.org

BirdWatch Ireland
*The leading voluntary conservation organisation
in Ireland, devoted to the conservation and
protection of Ireland's wild birds and their habitats.*
Rockingham House, Newcastle
Co. Wicklow, Ireland
Tel: (353) (01) 281 9878
www.birdwatchireland.ie

British Trust for Ornithology
*The BTO offers birdwatchers the opportunity to
learn more about birds by taking part in surveys
such as the Garden BirdWatch or the Nest
Record Scheme.*
The Nunnery, Thetford, Norfolk IP24 2PU
Tel: 01842 750 050
www.bto.org

C. J. Wildbird Foods Ltd
*C. J. Wildbird Foods is Britain's leading supplier
of birdfeeders and foodstuffs, via mail order. The
company produces a free handbook containing
advice on feeding garden birds.*
The Rea, Upton Magna, Shrewsbury SY4 4UR
Tel: 0800 731 2820 (Freephone)
www.birdfood.co.uk

Field Studies Council
*The FSC provides opportunities for people to
discover and understand the natural environment.*
Montford Bridge, Preston Montford, Shrewsbury,
SY4 1HW
Tel: 0845 345 4071
www.field-studies-council.org

Royal Society for the Protection of Birds (RSPB)
*Britain's leading bird conservation organization
with over one million members. It runs more
than 100 bird reserves up and down the country.*
The Lodge, Sandy, Bedfordshire SG19 2DL
Tel: 01767 680 551
www.rspb.org.uk

The Wildfowl & Wetlands Trust
*The WWT is dedicated to conservation of the
world's wetlands and their birds. It runs nine
centres in the UK, including the new Wetland
Centre in Barnes, West London, and the
famous Slimbridge site. It is supported by
over 139,000 members.*
Slimbridge, Gloucestershire GL2 7BT
Tel: 01453 890 333
www.wwt.org.uk

Wildsounds
*Britain's leading supplier of birdsong tapes
and CDs, including several on garden birds.*
Dept HTWB, Cross Street, Salthouse, Norfolk
NR25 7XH
Tel: 01263 741 100
www.wildsounds.com

Further Reading

MAGAZINES

BBC Wildlife
Available monthly from news-
agents, or by subscription from:
BBC Wildlife Subscriptions,
PO Box 279, Sittingbourne,
Kent ME9 8DF
Tel: 01795 414 718

Birdwatch
Available monthly from larger
newsagents, or by subscription
from:Warners, West Street,
Bourne, Lincolnshire
PE10 9PH
Tel: 01778 392 027

Bird Watching
Available monthly from larger
newsagents, or by subscription
from:Bretton Court,
Peterborough PE3 8DZ
Tel: 0845 601 1356

British Birds
Available monthly by subscrip-
tion only from:The Banks,
Mountfield, Robertsbridge,
East Sussex TN32 5JY
Tel: 01580 882 039

GENERAL BIRD BEHAVIOUR

**Bird Identification and
Fieldcraft**
Mark Ward
New Holland, 2005

**The Cambridge Encyclopaedia
of Ornithology**
Edited by Michael Brooke
and Tim Birkhead
Cambridge University Press,
1991

**The Complete Back Garden
Birdwatcher**
Dominic Couzens
New Holland, 2005

A Dictionary of Birds
Edited by Bruce Campbell and
Elizabeth Lack
Poyser, 1985

Understanding Bird Behaviour
Stephen Moss
New Holland, 2003

SPECIFIC BIRD BEHAVIOUR

The BTO Migration Atlas
John Marchant
Poyser, 2003

Bird Migration
Dominic Couzens
New Holland, 2004

Bird Migration
Robert Burton
Aurum, 1992

Bird Flight
Robert Burton
Facts on File, 1990

Birds and Berries
Barbara and David Snow
T & A. D. Poyser, 1983

Weather and Bird Behaviour
Norman Elkins
T & A.D. Poyser, 1983

Bird Song
C. K. Catchpole and
P. J. B. Slater
Cambridge University Press, 1985

**Bird Songs and Calls of Britain
and Northern Europe**
Geoff Sample
Collins, 1996

The Minds of Birds
Alexander Skutch
Texas A&M University Press,
1997

The Private Life of Birds
Stephen Moss
New Holland, 2006

GENERAL

**Attracting Birds to Your
Garden**
Stephen Moss &
David Cottridge
New Holland, 2000

Birds of Britain and Ireland
Bill Oddie
New Holland, 2002

**Birdwatcher's Pocket Field
Guide**
Mark Golley
New Holland, 2003

The Garden Bird Handbook
Stephen Moss
New Holland, 2003

How to Birdwatch
Stephen Moss
New Holland, 2003

**The Wildlife Trusts Guide
to Birds**
Series editor: Nicholas
Hammond
New Holland, 2002

Index

Acknowledgements

If this book relied solely on my own observations, it would be slender and very incomplete. For many years I wrote the Nature Note column in *The Daily Telegraph* and many readers wrote to tell me of their encounters with birds and other wildlife. These often raised questions that I could not answer. I followed these up by browsing in libraries or seeking the knowledge of experts in various fields of ornithology. I have also gathered information for this book from a number of published sources, including various ornithological journals and notably the monthly magazine, *British Birds*, and two admirable publications from the British Trust for Ornithology - *BTO News* and *Garden Birdwatch*.

I am grateful for permission to quote from the following copyright material:

The Robin by Walter de la Mare: The Literary Trustees of and the Society of Authors as their representative.

The Englishman's Year by H. J. Massingham: The Society of Authors as the Literary Representative of the Estate of H. J. Massingham.

The Truth by W. H. Davies: Mrs H. M. Davies Will Trust.

King Solomon's Ring by Konrad Lorenz, first published by Methuen & Co Ltd., 1952: Thomson Publishing Services.

The Publishers have made every effort to obtain permission for all quotations used in this book.

PHOTOGRAPHIC ACKNOWLEDGEMENTS:
All photographs by Steve Young except the following: p85, David Cotteridge; p102, Dennis Green.

ARTWORK ACKNOWLEDGEMENTS:
Richard Allen: 12, 14, 16, 17, 20, 21, 22, 23, 24, 26, 27, 31 (t, b), 33, 35(b), 37(t, b), 39(b), 42, 43 (t, b), 50(t), 52(b), 53, 54, 56(b), 65, 70(t, m, b), 71, 72, 76, 77, 79, 80(t, b), 81, 86(t, b), 87, 88, (t, b), 91, 93, 94, 95(tr, tl, bl, br), 96(t, b), 97, 98, 99(t, b), 103(t, b), 104, 105(b), 107(t, b), 108, 114, 117, 118(b), 120, 122, 124, 126, 128, 129(t, b), 131(t), 132, 133(tl, tr, b), 134, 138, 139(t,b). Clive Byers: 38, 131(b). Dave Daly: 4(t, b), 5(t, b), 18, 28(t, b), 34, 35(t), 39(t), 45(t, b), 47(t, b), 49, 50(b), 52(t), 56(t), 58(t, b), 63, 66, 67, 68, 73, 74(t, b), 75, 84, 89, 105(t), 109, 118(t), 135, 136, 137. The graphs on p116 are from The Wren by Edward A Armstrong, Collins, 1955.